C++ DESIGN PATTERNS AND DERIVATIVES PRICING

2nd edition

Design patterns are the cutting-edge paradigm for programming in object-oriented languages. Here they are discussed in the context of implementing financial models in C++. Assuming only a basic knowledge of C++ and mathematical finance, the reader is taught how to produce well-designed, structured, reusable code via concrete examples.

This new edition includes several new chapters describing how to increase robustness in the presence of exceptions, how to design a generic factory, how to interface C++ with EXCEL, and how to improve code design using the idea of decoupling. Complete ANSI/ISO compatible C++ source code is hosted on an accompanying website for the reader to study in detail, and reuse as they see fit.

A good understanding of C++ design is a necessity for working financial mathematicians; this book provides a thorough introduction to the topic.

Mathematics, Finance and Risk

Editorial Board

C++ DESIGN PATTERNS AND DERIVATIVES PRICING

M. S. JOSHI

University of Melbourne

CAMBRIDGE UNIVERSITY PRESS
Cambridge, New York, Melbourne, Madrid, Cape Town, Singapore, São Paulo, Delhi

Cambridge University Press
The Edinburgh Building, Cambridge CB2 8RU, UK

Published in the United States of America by Cambridge University Press, New York

www.cambridge.org
Information on this title: www.cambridge.org/9780521721622

First published 2008

Printed in the United Kingdom at the University Press, Cambridge

A catalogue record for this publication is available from the British Library

ISBN 978-0-521-72162-2 paperback

To Jane

Contents

Preface

This book is aimed at a reader who has studied an introductory book on mathematical finance and an introductory book on C++ but does not know how to put the two together. My objective is to teach the reader not just how to implement models in C++ but more importantly how to think in an object-oriented way. There are already many books on object-oriented programming; however, the examples tend not to feel real to the financial mathematician so in this book we work exclusively with examples from derivatives pricing.

We do not attempt to cover all sorts of financial models but instead examine a few in depth with the objective at all times of using them to illustrate certain OO ideas. We proceed largely by example, rewriting, our designs as new concepts are introduced, instead of working out a great design at the start. Whilst this approach is not optimal from a design standpoint, it is more pedagogically accessible. An aspect of this is that our examples are designed to emphasize design principles rather than to illustrate other features of coding, such as numerical efficiency or exception safety.

We commence by introducing a simple Monte Carlo model which does not use OO techniques but rather is the simplest procedural model for pricing a call option one could write. We examine its shortcomings and discuss how classes naturally arise from the concepts involved in its construction.

In Chapter 2, we move on to the concept of encapsulation – the idea that a class allows to express a real-world analogue and its behaviours precisely. In order to illustrate encapsulation, we look at how a class can be defined for the pay-off of a vanilla option. We also see that the class we have defined has certain defects, and this naturally leads on to the open–closed principle.

In Chapter 3, we see how a better pay-off class can be defined by using inheritance and virtual functions. This raises technical issues involving destruction and passing arguments, which we address. We also see how this approach is compatible with the open–closed principle.

Using virtual functions causes problems regarding the copying of objects of unknown type, and in Chapter 4 we address these problems. We do so by introducing virtual constructors and the bridge pattern. We digress to discuss the 'rule of three' and the slowness of new. The ideas are illustrated via a vanilla options class and a parameters class.

With these new techniques at our disposal, we move on to looking at more complicated design patterns in Chapter 5. We first introduce the strategy pattern that expresses the idea that decisions on part of an algorithm can be deferred by delegating responsibilities to an auxiliary class. We then look at how templates can be used to write a wrapper class that removes a lot of our difficulties with memory handling. As an application of these techniques, we develop a convergence table using the decorator pattern.

In Chapter 6, we look at how to develop a random numbers class. We first examine why we need a class and then develop a simple implementation which provides a reusable interface and an adequate random number generator. We use the implementation to introduce and illustrate the adapter pattern, and to examine further the decorator pattern.

We move on to our first non-trivial application in Chapter 7, where we use the classes developed so far in the implementation of a Monte Carlo pricer for path-dependent exotic derivatives. As part of this design, we introduce and use the template pattern. We finish with the pricing of Asian options.

We shift from Monte Carlo to trees in Chapter 8. We see the similarities and differences between the two techniques, and implement a reusable design. As part of the design, we reuse some of the classes developed earlier for Monte Carlo.

We return to the topic of templates in Chapter 9. We illustrate their use by designing reusable solver classes. These classes are then used to define implied volatility functions. En route, we look at function objects and pointers to member functions. We finish with a discussion of the pros and cons of templatization.

In Chapter 10, we look at our most advanced topic: the factory pattern. This patterns allows the addition of new functionality to a program without changing any existing files. As part of the design, we introduce the singleton pattern.

We pause in Chapter 11 to classify, summarize, and discuss the design patterns we have introduced. In particular, we see how they can be divided into creational, structural, and behavioural patterns. We also review the literature on design patterns to give the reader a guide for further study.

The final four chapters are new for the second edition. In these our focus is different: rather than focussing exclusively on design patterns, we look at some other important aspects of good coding that neophytes to C++ tend to be unaware of.

In Chapter 12, we take a historical look at the situation in 2007 and at what has changed in recent years both in C++ and the financial engineering community's use of it.

The study of exception safety is the topic of Chapter 13. We see how making the requirement that code functions well in the presence of exceptions places a large number of constraints on style. We introduce some easy techniques to deal with these constraints.

In Chapter 14, we return to the factory pattern. The original factory pattern required us to write similar code every time we introduced a new class hierarchy; we now see how, by using argument lists and templates, a fully general factory class can be coded and reused forever.

In Chapter 15, we look at something rather different that is very important in day-to-day work for a quant: interfacing with EXCEL. In particular, we examine the xlw package for building xlls. This package contains all the code necessary to expose a C++ function to EXCEL, and even contains a parser to write the new code required for each function.

The concept of physical design is introduced in Chapter 16. We see how the objective of reducing compile times can affect our code organization and design.

The code for the examples in the first 11 chapters of this book can be freely downloaded from www.markjoshi.com/design, and any bugfixes will be posted there. The code for the remaining chapters is taken from the xlw project and can be downloaded from xlw.sourceforge.net. All example code is taken from release 2.1.

Acknowledgements

I am grateful to the Royal Bank of Scotland for providing a stimulating environment in which to learn, study and do mathematical finance. Most of my views on coding C++ and financial modelling have been developed during my time working there. My understanding of the topic has been formed through daily discussions with current and former colleagues including Chris Hunter, Peter Jäckel, Dherminder Kainth, Sukhdeep Mahal, Robin Nicholson and Jochen Theis. I am also grateful to a host of people for their many comments on the manuscript, including Alex Barnard, Dherminder Kainth, Rob Kitching, Sukhdeep Mahal, Nadim Mahassen, Hugh McBride, Alan Stacey and Patrik Sundberg. I would also like to thank David Tranah and the rest of the team at Cambridge University Press for their careful work and attention to detail. Finally my wife has been very supportive.

I am grateful to a number of people for their comments on the second edition, with particular thanks to Chris Beveridge, Narinder Claire, Nick Denson and Lorenzo Liesch.

1

A simple Monte Carlo model

1.1 Introduction

In the first part of this book, we shall study the pricing of derivatives using Monte Carlo simulation. We do this not to study the intricacies of Monte Carlo but because it provides many convenient examples of concepts that can be abstracted. We proceed by example, that is we first give a simple program, discuss its good points, its shortcomings, various ways round them and then move on to a new example. We carry out this procedure repeatedly and eventually end up with a fancy program. We begin with a routine to price vanilla call options by Monte Carlo.

1.2 The theory

We commence by discussing the theory. The model for stock price evolution is

$$dS_t = \mu S_t dt + \sigma S_t dW_t, \tag{1.1}$$

and a riskless bond, B, grows at a continuously compounding rate r. The Black–Scholes pricing theory then tells us that the price of a vanilla option, with expiry T and pay-off f, is equal to

$$e^{-rT}\mathbb{E}(f(S_T)),$$

where the expectation is taken under the associated risk-neutral process,

$$dS_t = r S_t dt + \sigma S_t dW_t. \tag{1.2}$$

We solve equation (1.2) by passing to the log and using Ito's lemma; we compute

$$d \log S_t = \left(r - \frac{1}{2}\sigma^2\right) dt + \sigma dW_t. \tag{1.3}$$

As this process is constant-coefficient, it has the solution

$$\log S_t = \log S_0 + \left(r - \frac{1}{2}\sigma^2\right) t + \sigma W_t. \tag{1.4}$$

1

Since W_t is a Brownian motion, W_T is distributed as a Gaussian with mean zero and variance T, so we can write

$$W_T = \sqrt{T}N(0, 1), \tag{1.5}$$

and hence

$$\log S_T = \log S_0 + \left(r - \frac{1}{2}\sigma^2 \right) T + \sigma\sqrt{T}N(0, 1), \tag{1.6}$$

or equivalently,

$$S_T = S_0 e^{(r-\frac{1}{2}\sigma^2)T + \sigma\sqrt{T}N(0,1)}. \tag{1.7}$$

The price of a vanilla option is therefore equal to

$$e^{-rT}\mathbb{E}\left(f\left(S_0 e^{(r-\frac{1}{2}\sigma^2)T + \sigma\sqrt{T}N(0,1)} \right) \right).$$

The objective of our Monte Carlo simulation is to approximate this expectation by using the law of large numbers, which tells us that if Y_j are a sequence of identically distributed independent random variables, then with probability 1 the sequence

$$\frac{1}{N}\sum_{j=1}^{N} Y_j$$

converges to $\mathbb{E}(Y_1)$.

So the algorithm to price a call option by Monte Carlo is clear. We draw a random variable, x, from an $N(0, 1)$ distribution and compute

$$f\left(S_0 e^{(r-\frac{1}{2}\sigma^2)T + \sigma\sqrt{T}x} \right),$$

where $f(S) = (S - K)_+$. We do this many times and take the average. We then multiply this average by e^{-rT} and we are done.

1.3 A simple implementation of a Monte Carlo call option pricer

A first implementation is given in the program `SimpleMCMain1.cpp`.

Listing 1.1 (`SimpleMCMain1.cpp`)

```
//  requires Random1.cpp


#include <Random1.h>
#include <iostream>
#include <cmath>
using namespace std;
```

```cpp
double SimpleMonteCarlo1(double Expiry,
                        double Strike,
                        double Spot,
                        double Vol,
                        double r,
                        unsigned long NumberOfPaths)
{
    double variance = Vol*Vol*Expiry;
    double rootVariance = sqrt(variance);
    double itoCorrection = -0.5*variance;

    double movedSpot = Spot*exp(r*Expiry +itoCorrection);
    double thisSpot;
    double runningSum=0;

    for (unsigned long i=0; i < NumberOfPaths; i++)
    {
    double thisGaussian = GetOneGaussianByBoxMuller();
    thisSpot = movedSpot*exp( rootVariance*thisGaussian);
    double thisPayoff = thisSpot - Strike;
    thisPayoff = thisPayoff >0 ? thisPayoff : 0;
    runningSum += thisPayoff;
    }

    double mean = runningSum / NumberOfPaths;
    mean *= exp(-r*Expiry);
    return mean;
}

int main()
{
    double Expiry;
    double Strike;
    double Spot;
    double Vol;
    double r;
    unsigned long NumberOfPaths;
    cout << "\nEnter expiry\n";
    cin >> Expiry;
```

```
    cout << "\nEnter strike\n";
    cin >> Strike;

    cout << "\nEnter spot\n";
    cin >> Spot;

    cout << "\nEnter vol\n";
    cin >> Vol;

    cout << "\nr\n";
    cin >> r;

    cout << "\nNumber of paths\n";
    cin >> NumberOfPaths;

    double result = SimpleMonteCarlo1(Expiry,
                                      Strike,
                                      Spot,
                                      Vol,
                                      r,
                                      NumberOfPaths);

    cout <<"the price is " << result << "\n";

    double tmp;
    cin >> tmp;

 return 0;
}
```

Our program uses the auxiliary files Random1.h and Random1.cpp.

Listing 1.2 (Random1.h)

```
#ifndef RANDOM1_H
#define RANDOM1_H

double GetOneGaussianBySummation();
double GetOneGaussianByBoxMuller();
#endif
```

Listing 1.3 (Random1.cpp)

```cpp
#include <Random1.h>
#include <cstdlib>
#include <cmath>

// the basic math functions should be in namespace
// std but aren't in VCPP6
#if !defined(_MSC_VER)
using namespace std;
#endif

double GetOneGaussianBySummation()
{
double result=0;

for (unsigned long j=0; j < 12; j++)
result += rand()/static_cast<double>(RAND_MAX);

result -= 6.0;

return result;
}

double GetOneGaussianByBoxMuller()
{
double result;

double x;
double y;

double sizeSquared;
do
{
x = 2.0*rand()/static_cast<double>(RAND_MAX)-1;
y = 2.0*rand()/static_cast<double>(RAND_MAX)-1;
sizeSquared = x*x + y*y;
}
while
( sizeSquared >= 1.0);
```

$$\sum_{i=1}^{12} x_i - 6$$

```
result = x*sqrt(-2*log(sizeSquared)/sizeSquared);

return result;
}
```

We first include the header file Random1.h. Note that the program has <Random1.h> rather than "Random1.h". This means that we have set our compiler settings to look for header files in the directory where Random1.h is. In this case, this is in the directory C/include. (In Visual C++, the directories for include files can be changed via the menus tools, options, directories.)

Random1.h tells the main file that the functions

```
double GetOneGaussianBySummation()
```

and

```
double GetOneGaussianByBoxMuller()
```

exist. We include the system file iostream as we want to use cin and cout for the user interface. The system file cmath is included as it contains the basic mathematical functions exp and sqrt.

We have the command using namespace std because all the standard library commands are contained in the namespace std. If we did not give the using directive, then we would have to prefix all their uses by std::, so then it would be std::cout rather than cout.

The function SimpleMonteCarlo1 does all the work. It takes in all the standard inputs for the Black–Scholes model, the expiry and strike of the option, and in addition the number of paths to be used in the Monte Carlo.

Before starting the Monte Carlo we precompute as much as possible. Thus we compute the variance of the log of the stock over the option's life, the adjustment term $-\frac{1}{2}\sigma^2 T$ for the drift of the log, and the square root of the variance. Whilst we cannot precompute the final value of spot, we precompute what we can and put it in the variable movedSpot.

We initialize the variable, runningSum, to zero as it will store the sum so far of the option pay-offs at all times.

We now loop over all the paths. For each path, we first draw the random number from the $N(0, 1)$ distribution using the Box–Muller algorithm and put it in the variable thisGaussian.

The spot at the end of the path is then computed and placed in thisSpot. Note that although our derivation of the SDE involved working with the log of the spot, we have carefully avoided using log in this routine. The reason is that log and exp are slow to compute in comparison to addition and multiplication, we therefore want to make as few calls to them as possible.

We then compute the call option's pay-off by subtracting the strike and taking the maximum with zero. The pay-off is then added to `runningSum` and the loop continues.

Once the loop is complete, we divide by the number of paths to get the expectation. Finally, we discount to get our estimate of the price which we return.

The `main` program takes in the inputs from the user, calls the Monte Carlo function, and displays the results. It asks for a final input to stop the routine from returning before the user has had a chance to read the results.

1.4 Critiquing the simple Monte Carlo routine

The routine we have written runs quickly and does what it was intended to do. It is a simple straightforward procedural program that performs as required. However, if we worked with this program we would swiftly run into annoyances. The essence of good coding is reusability. What does this mean? One simple definition is that code is reusable if someone has reused it. Thus reusability is as much a social concept as a technical one. What will make it easy for someone to reuse your code? Ultimately, the important attributes are clarity and elegance of design. If another coder decides that it would take as much effort to recode your routines as to understand them, then he will recode, and his inclination will be to recode in any case, as it is more fun than poring over someone else's implementation.

The second issue of elegance is equally important. If the code is clear but difficult to adapt then another coder will simply abandon it, rather than put lots of effort into forcing it to work in a way that is unnatural for how it was built.

The demands of reusability therefore mean we should strive for clarity and elegance. In addition, we should keep in mind when considering our original design the possibility that in future our code might need to be extended.

We return to our simple Monte Carlo program. Suppose we have a boss and each day he comes by and asks for our program to do something more. If we have designed it well then we will simply have to add features; if we have designed poorly then we will have to rewrite existing code.

So what might the evil boss demand?

"Do puts as well as calls!"

"I can't see how accurate the price is, put in the standard error."

"The convergence is too slow, put in anti-thetic sampling."

"I want the most accurate price possible by 9am tomorrow so set it running for 14 hours."

"It's crucial that the standard error is less than 0.0001, so run it until that's achieved. We're in a hurry though so don't run it any longer than strictly necessary."

"I read about low-discrepancy numbers at the weekend. Just plug them in and see how good they are."

"Apparently, standard error is a poor measure of error for low-discrepancy simulations. Put in a convergence table instead."

"Hmm, I miss the standard error can we see that too."

"We need a digital call pricer now!"

"What about geometric average Asian calls?"

"How about arithmetic average Asian puts?"

"Take care of variable parameters for the volatility and interest rates."

"Use the geometric Asian option as a control variate for the arithmetic one."

"These low-discrepancy numbers apparently only work well if you Brownian bridge. Put that in as well."

"Put in a double digital geometric Asian option."

"What about jump risk? Put in a jump-diffusion model."

To adapt the routine as written would require a rewrite to do any of these. We have written the simplest routine we could think of, without considering design issues. This means that each change is not particularly natural and requires extra work.

For example, with this style of programming how would we would do the put option?

Option one: copy the function, change the name by adding put at the end, and rewrite the two lines where the pay-off is computed.

Option two: pass in an extra parameter, possibly as an `enum` and compute the pay-off via a `switch` statement in each loop of the Monte Carlo. The problem with the first option is that when we come to the next task, we have to adapt both the functions in the same way and do the same thing twice. If we then need more pay-offs this will rapidly become a maintenance nightmare.

The issues with the other option are more subtle. One problem is that a switch statement is an additional overhead so that the routine will now run a little slower. A deeper problem is that when we come to do a different routine which also uses a pay-off, we will have to copy the code from inside the first routine or rewrite it as necessary. This once again becomes a maintenance problem; every time we want to add a new sort of pay-off we would have to go through every place where pay-offs are used and add it on.

A C style approach to this problem would be to use a function pointer, we pass a pointer to a function as an argument to the Monte Carlo. The function pointed to is then called via the pointer in each loop to specify the price. Note that the call to the function would have to specify the strike as well as spot since the function could not know its value. Note also that if we wanted to do a double-digital option we would have problems as the double digital pays if and only if spot is between two levels, and we only have one argument, the strike, to play with.

The C++ approach to this problem is to use a class. The class would encapsulate the behaviour of the pay-off of a vanilla option. A pay-off object would then be passed into the function as an argument and in each loop a method expressing its value would be called to output the price for that pay-off. We look at the implementation of such a class in the next chapter.

1.5 Identifying the classes

In the previous section, we saw that the problem of wanting to add different sorts of vanilla options led naturally to the use of a class to encapsulate the notion of a pay-off. In this section, we look at identifying other natural classes which arise from the boss's demands.

Some of the demands were linked to differing forms that the boss wanted the information in. We could therefore abstract this notion by creating a statistics gatherer class.

We also had differing ways of terminating the Monte Carlo. We could terminate on time, on standard error or simply after a fixed number of paths. We could abstract this by writing a terminator class.

There were many different issues with the method of random number generation. The routine as it stands relies on the inbuilt generator which we do not know much about. We therefore want to be able to use other random number generators. We also want the flexibility of using low-discrepancy numbers which means another form of generation. (In addition, Box–Muller does not work well with low-discrepancy numbers so we will need flexibility in the inputs.) Another natural abstraction is therefore a random number generator class.

As long as our option is vanilla then specifying its parameters via pay-off and strike is fairly natural and easy; however, it would be neater to have one class that contains both pieces of information. More generally, when we pass to path-dependent exotic options, it becomes natural to have a class that expresses the option's properties. What would we expect such a class to do? Ultimately, an easy way to decide what the class should and should not know is to think of whether a piece of information would be contained in the term-sheet. Thus the class would know the pay-off of the option. It would know the expiry time. If it was an Asian it would know the averaging dates. It would also know whether the averaging was geometric or arithmetic. It would not know anything about interest rates, nor the value of spot nor the volatility of the spot rate as none these facts are contained in the term-sheet. The point here is that by choosing a real-world concept to encapsulate, it is easy to decide what to include or not to include. It is also easy for another user to understand how you have decided what to include or not to include.

What other concepts can we identify? The concept of a variable parameter could be made into a class. The process from which spot is drawn is another concept. The variable interest rates could be encapsulated via a class that expresses the notion of a discount curve.

1.6 What will the classes buy us?

Suppose that having identified all these classes, we implement them. What do we gain?

The first gain is that because these classes encapsulate natural financial concepts, we will need them when doing other pieces of coding. For example, if we have a class that does yield curves then we will use it time and time again, as to price any derivative using any reasonable method involves knowledge of the discount curve. Not only will we save time on the writing of code but we will also save time on the debugging. A class that has been tested thoroughly once has been tested forever and in addition, any little quirks that evade the testing regime will be found through repeated reuse. The more times and ways something has been reused the fewer the bugs that will be left. So using reusable classes leads to more reliable code as well as saving us coding time. Debugging often takes at least as much time as coding in any case, so saving time on debugging is a big benefit.

A second gain is that our code becomes clearer. We have written the code in terms of natural concepts, so another coder can identify the natural concepts and pick up our code much more easily.

A third gain is that the classes will allow us to separate interface from implementation. All the user needs to know about a pay-off class or discount curve class are what inputs yield what outputs? How the class works internally does not matter. This has multiple advantages. The first is that the class can be reused without the coder having to study its internal workings. The second advantage is that because the defining characteristic of the class is what it does but not how it does it, we can change how it does it at will. And crucially, we can do this without rewriting the rest of our program. One aspect of this is that we can first quickly write a suboptimal implementation and improve it later at no cost. This allows us to provide enough functionality to test the rest of the code before devoting a lot of time to the class. A third advantage of separating interface from implementation is that we can write multiple classes that implement the same interface and use them without rewriting all the interface routines. This is one of the biggest advantages of object-oriented design.

In the next chapter, we look at some of these concepts in the concrete case of a pay-off class.

1.7 Why object-oriented programming?

This is a book about implementing pricing models using object-oriented C++ programs. The reader may ask why this is worth learning. A short answer is that this is the skill you need if you want a job working as a quantitative analyst or quantitative developer. But this begs the question of why this is the required skill.

Object-oriented programming has become popular as computer projects have become larger and larger. A single project may now involve millions of lines of code. No single programmer will ever be able to hold all of that code in his mind at once. Object-oriented programming provides us with a way of coding that corresponds to natural mental maps. We know what each class of objects does, and more importantly we tightly define how they can interact with each other. This allows a clear map in the coder's mind of how the code fits together. And equally importantly, this allows easy communication of the code's structure to other programmers in the team.

When the coder needs to focus in on a particular part of the code, he need only look at the interior of the particular object involved and its interface with other objects. As long as the interface is not broken, and the new object lives up to the same responsibilities as the old one then there is no danger of unexpected ramifications (i.e. bugs) in distant parts of the code. Thus object-oriented programming leads to more robust code that is easier for teams to work on.

1.8 Key points

In this chapter, we have looked at how to implement a simple Monte Carlo routine on a procedural program. We then criticized it from the point of view of easy extensibility and reuse.

- Options can be priced by risk-neutral expectation.
- Monte Carlo uses the Law of Large Numbers to approximate this risk-neutral expectation.
- Reuse is as much a social issue as a technical one.
- Procedural programs can be hard to extend and reuse.
- Classes allow us to encapsulate concepts which makes reuse and extensibility a lot easier.
- Making classes closely model real-world concepts makes them easier to design and to explain.

- Classes allow us to separate the design of the interface from the coding of the implementation.

1.9 Exercises

Exercise 1.1 Modify the program given to price puts.

Exercise 1.2 Modify the program given to price double digitals.

Exercise 1.3 Change the program so that the user inputs a string which specifies the option pay-off.

Exercise 1.4 Identify as many classes as you can in the evil boss's list of demands.

2

Encapsulation

2.1 Implementing the pay-off class

In the last chapter, we looked at a simple Monte Carlo pricer and concluded that the program would be improved if we used a class to encapsulate the notion of the pay-off of a vanilla option. In this section, we look at how such a pay-off might be implemented. In the files PayOff1.h and Payoff1.cpp, we develop such a class. Before looking at the source file, PayOff1.cpp, we examine the more important header file. The header file is much more important because it is all that other files will see, and ideally it is all that the other files need to see.

Listing 2.1 (PayOff1.h)

```
#ifndef PAYOFF_H
#define PAYOFF_H

class PayOff
{
public:
    enum  OptionType {call, put};
    PayOff(double Strike_, OptionType TheOptionsType_);
    double operator()(double Spot) const;

private:
    double Strike;
    OptionType TheOptionsType;
};
#endif
```

We use an enum to distinguish between different sorts of pay-offs. If we ever want more than put and call, we would add them to the list. We will discuss

13

more sophisticated approaches in Chapter 3 but this approach is enough to get us started.

The constructor

```
PayOff(double Strike_, OptionType TheOptionsType_)
```

takes in the strike of the option and the type of the option pay-off.

The main method of the class is

```
double PayOff::operator()(double spot) const
```

The purpose of this method is that given a value of spot, it returns the value of the pay-off.

Note the slightly odd syntax: we have overloaded `operator()`. To call this method for an object `thePayoff` with spot given by S we would write `thePayoff(S)`. Thus when use the object it appears more like a function than an object; such objects are often called function objects or functors. (Note that the C++ concept of functor is quite different from the pure mathematical one.)

We have defined `operator()` to be `const`. This means that the method does not modify the state of the object. This is as we would expect: computing the pay-off does not change the strike or the type of an option.

Suppose we did not specify that `operator()` was `const`; what would happen? The same functionality would be provided and the code would still compile. However, if a pay-off object was declared `const` at some point by a user then the compiler would complain if we tried to call `operator()`, and give us a complicated message to the effect that we cannot call non `const` methods of `const` objects. Thus if we want a method to be usable in `const` objects we must declare the method `const`.

An alternative approach, adopted by some programmers, is not to use `const` anywhere. The argument goes along the lines of "If we use `const` in some places, we must use it everywhere, and all it does is cause trouble and stop me doing what I want so why bother? Life will be much easier if we just do not use it."

Yet I use `const` as much as possible. The reason is that it is a safety enforcement mechanism. Thinking about `const` forces me to consider the context of an object and whether or not I wish it to change when doing certain things. If I am woolly in my thinking then the compiler will generally squeal when I attempt to compile, and thus errors are picked up at compile time instead of at run time.

A second benefit is that by giving the compiler the extra information that objects are `const`, we allow it to make extra optimizations. Code on a good compiler can therefore run faster when `const` is used. Now we are ready for `PayOff1.cpp`.

Listing 2.2 (`PayOff1.cpp`)

```
#include <PayOff1.h>
#include <MinMax.h>

PayOff::PayOff(double Strike_, OptionType TheOptionsType_)
:
    Strike(Strike_), TheOptionsType(TheOptionsType_)
{
}

double PayOff::operator ()(double spot) const
{
    switch (TheOptionsType)
    {
    case call :
        return max(spot-Strike,0.0);

    case put:
        return max(Strike-spot,0.0);

    default:
        throw("unknown option type found.");
    }
}
```

2.2 Privacy

We have declared the data in the pay-off class to be private. This means that the data cannot be accessed by code outside the class. The only code that can see, let alone change, their values are the constructor and the method `operator()`. What does this buy us? After all, for some reason the user might want to know the strike of an option and we have denied him the possibility of finding it out.

The reason is that as soon we let the user access the data directly, it is much harder for us to change how the class works. We can think of the class as a contract between coder and user. We, the coder, have provided the user with an interface: if he gives us spot we will give him the value of the pay-off. This is all we have contracted to provide. The user therefore expects and receives precisely that and no more.

For example, if the user could see the strike directly and accessed it, and if we changed the class so that the strike was no longer stored directly (which we shall do in the next chapter), then we would get compile errors from everywhere the strike was accessed. If the class had been used a lot in many files, in many different projects (which is after all the objective of reuse), then to find every place where strike had been accessed would be a daunting task. In fact, if this were the case we would probably consider finding them all a considerable waste of time, and we would probably give up reforming the internal workings.

Thus by making the data private, we can enforce the contract between coder and user in such a way that the contract does not say anything about the interior workings of the class. If for some reason, we wanted the user to be able to know the strike of the pay-off then we could add in an extra method `double GetStrike() const`. Whilst this would seem to undo all the benefits of using private data, this is not so since it still allows us the possibility of storing the data in a totally different way.

To conclude, by using private data we can rework the interior workings of a class as and when we need to without worrying about the impact on other code. That is, private data helps us to separate interface and implementation.

2.3 Using the pay-off class

Having designed a pay-off class, we want to use it in our Monte Carlo routines. We do this in the function `SimpleMonteCarlo2` which is declared in `SimpleMC.h` and defined in `SimpleMC.cpp`.

Listing 2.3 (`SimpleMC.h`)

```
#ifndef SIMPLEMC_H
#define SIMPLEMC_H
#include <PayOff1.h>

double SimpleMonteCarlo2(const PayOff& thePayOff,
                         double Expiry,
                         double Spot,
                         double Vol,
                         double r,
                         unsigned long NumberOfPaths);
#endif
```

Listing 2.4 (`SimpleMC.cpp`)

```
#include <SimpleMC.h>
#include <Random1.h>
#include <cmath>
```

```
// the basic math functions should be in
// namespace std but aren't in VCPP6
#if !defined(_MSC_VER)
using namespace std;
#endif

double SimpleMonteCarlo2(const PayOff& thePayOff,
                         double Expiry,
                         double Spot,
                         double Vol,
                         double r,
                         unsigned long NumberOfPaths)
{
    double variance = Vol*Vol*Expiry;
    double rootVariance = sqrt(variance);
    double itoCorrection = -0.5*variance;

    double movedSpot = Spot*exp(r*Expiry +itoCorrection);
    double thisSpot;
    double runningSum=0;

    for (unsigned long i=0; i < NumberOfPaths; i++)
    {
        double thisGaussian = GetOneGaussianByBoxMuller();
        thisSpot = movedSpot*exp( rootVariance*thisGaussian);
        double thisPayOff = thePayOff(thisSpot);
        runningSum += thisPayOff;
    }
    double mean = runningSum / NumberOfPaths;
    mean *= exp(-r*Expiry);
    return mean;
}
```

Handwritten annotations: $\sigma^2 T$, $\sigma\sqrt{T}$, $-\frac{1}{2}\sigma^2 T$, $S e^{\left(r-\frac{1}{2}\sigma^2\right)T}$, $e^{\sigma\sqrt{T}\, Z}$

The main change from our original Monte Carlo routine is that we now input a PayOff object instead of a strike. The strike is, of course, now hidden inside the inputted object. We pass the object by reference and make it const as we have no need to change it inside our routine. Note that the only line of our algorithm that is new is

```
    double thisPayOff = thePayOff(thisSpot);
```

We illustrate how the routine might be called in `SimpleMCMain2.cpp`. Here we define both call and put objects and this shows that the routine can now be used without change to prices both types of options.

Listing 2.5 (`SimpleMCMain2.cpp`)

```
/*

requires

        PayOff1.cpp
        Random1.cpp,
        SimpleMC.cpp,
*/


#include<SimpleMC.h>
#include<iostream>
using namespace std;

int main()
{
        double Expiry;
        double Strike;
        double Spot;
        double Vol;
        double r;
        unsigned long NumberOfPaths;

        cout << "\nEnter expiry\n";
        cin >> Expiry;

        cout << "\nEnter strike\n";
        cin >> Strike;

        cout << "\nEnter spot\n";
        cin >> Spot;

        cout << "\nEnter vol\n";
        cin >> Vol;

        cout << "\nr\n";
        cin >> r;
```

```
cout << "\nNumber of paths\n";
cin >> NumberOfPaths;

PayOff callPayOff(Strike, PayOff::call);
PayOff putPayOff(Strike, PayOff::put);

double resultCall = SimpleMonteCarlo2(callPayOff,
                                    Expiry,
                                    Spot,
                                    Vol,
                                    r,
                                    NumberOfPaths);

double resultPut = SimpleMonteCarlo2(putPayOff,
                                    Expiry,
                                    Spot,
                                    Vol,
                                    r,
                                    NumberOfPaths);

cout <<"the prices are " << resultCall
                        << "  for the call and "
                        << resultPut
                        << " for the put\n";

double tmp;
cin >> tmp;

return 0;
}
```

2.4 Further extensibility defects

We have now set-up a pay-off class. One of our original objectives was to be able to extend the code easily without needing to rewrite other code when we add new pay-offs. Have we achieved this? Yes.

Suppose we now want to add in the digital call pay-off. What do we need to do? We add in an extra type, `digitalcall`, in the enum. We also change the `switch` statement in the `operator()` to include the extra case of a digital call and in that case return 1 if spot is above strike and 0 otherwise.

Everywhere the `PayOff` class has been used, in our case the Monte Carlo routine but possibly in many places, the digital call is now available just by passing in a `PayOff` object in an appropriate state.

Do this change for yourself and hit 'build'. One slightly non-obvious effect is that all the files which involve pay-offs are now recompiled. So although there is no obvious difference in those files, we see something slightly unexpected: as they include the file `PayOff1.h`, and it has changed, they must be recompiled too.

This is a warning sign that the code is not as independent as we would like. In an extreme case, we might not have access to source code for purchased libraries; any programming model that required us to recompile those libraries would be useless to us. In addition even if we could recompile, it might be a time-consuming process which we would prefer to avoid.

We would therefore like a way of programming that allows us not just to add functionality without modifying dependent files, but also to be able to do so without having to recompile existing files.

In fact, any solution that allowed us to do so would have to allow us to add an extra sort of pay-off without changing the pay-off class that the files using pay-offs see. For if the part that they see changes in any way, the compiler will insist on recompiling them even if the changes do not appear to be material.

2.5 The open–closed principle

The previous section's discussion leads naturally to a programming idea known as the *open-closed principle*. The 'open' refers to the idea that code should always be open for extension. So in this particular case, we should always be able to add extra pay-offs.

The 'closed' means that the file is 'closed for modification'. This refers to the idea that we should be able to do the extension without modifying any existing code, and we should be able to do so without even changing anything in any of the existing files.

This may seem a little counterintuitive; how can one possibly make new forms of pay-offs without changing the pay-off class? To illustrate the idea before presenting the C++ solution, we consider how we might do this using C style techniques.

Suppose instead of making the class contain an `enum` that defines the option type, we instead use a function pointer. Thus we replace `OptionType` with

```
double (*FunctionPtr)(double,double).
```

The constructor for the pay-off would then take in the strike and a function pointer. It would store both and when `operator()` was called it would dereference the pointer and call the function pointed to with spot and strike as its arguments.

This code achieves a lot of our objectives. We can put the function pointed to in a new file, so the existing file for the pay-off class does not change each time we add a new form of pay-off. This means that neither the pay-off file nor the Monte Carlo file which includes the header file for pay-off need to be recompiled let alone changed.

However, there is still a fly in the ointment. What do we do when the boss comes by and demands a double-digital pay-off? The double digital requires two strikes (or barrier levels) and we only have one slot, the strike. Now this was also a problem with our original pay-off class; it too had only one slot for the strike.

One solution would be to use an array to specify and store parameters. The strike would be replaced by an array in both constructor and class members. The function pointer would take the array and spot as it arguments. This could be made to work. However, the code is now starting to lose the properties of clarity and elegance which were a large part of our original objectives.

So although the function pointer gets us a long way it does not get us far enough and we need a more sophisticated approach. The issue is that we need a way of specifying an object for the pay-off which is not of a predetermined type. This object will contain all the data it needs know and no more. So for a vanilla call or put the object would contain the strike, but for a double digital it would contain both of the two barrier levels.

Fortunately, C++ was designed with just this sort of problem in mind, and in the next chapter we shall study how to use inheritance and virtual functions to overcome these problems. We have used the open–closed principle as a way of motivating the introduction of these concepts, and we shall repeatedly return to it as a paradigm for evaluating programming approaches.

2.6 Key points

In this chapter, we looked at one way of using a class to encapsulate the notion of a pay-off. We then saw some of the advantages of using such a class. We also saw that the class had not achieved all of our objectives.

- Using a pay-off class allows us to add extra forms of pay-offs without modifying our Monte Carlo routine.
- By overloading `operator()` we can make an object look like a function.
- `const` enforces extra discipline by forcing the coder to be aware of which code is allowed to change things and which code cannot.
- `const` code can run faster.

- The open-closed principle says that code should be open for extension but closed for modification.
- Private data helps us to separate interface from implementation.

2.7 Exercises

Exercise 2.1 Modify the pay-off class so that it can handle digital options.

Exercise 2.2 Modify the pay-off class so that it can handle double-digital options.

Exercise 2.3 Test whether on your compiler using `const` speeds code up.

3

Inheritance and virtual functions

3.1 'is a'

We reconsider our `PayOff` class. At the end of the last chapter, we concluded that we would like to be able to use differing sorts of objects as pay-offs without changing any of the code that takes in pay-off objects. In other words, we would like to be able to say that a call option pay-off 'is a' pay-off, and express this idea in code.

The mechanism for expressing the 'is a' relationship in C++ is inheritance. There are plenty of examples we have already seen where such 'is a' relationships are natural. For example, a call option is a vanilla option. The compiler's built-in random number generator is a random number generator. Box–Muller is a method of turning uniform random variables into Gaussian random variables. An Asian option is a path-dependent exotic option. An arithmetic Asian call option is an Asian option, as is a geometric Asian put option. Returning the mean is a way of gathering statistics for a Monte Carlo simulation. Specifying the continuously compounding rate is a way to specify the discount curve. The BlackScholes model is a model of stock price evolution.

Thus 'is a' relationships are very natural in mathematical finance (as well as in coding and life in general.) We shall use inheritance to express such relationships. We shall refer to the class we inherit from as the *base class* and the class which does the inheriting will be called the *inherited class*. The base class will always express a more general concept than the inherited class.

Note that there is nothing to stop us inheriting several times. For example, our base might be `PathDependentExoticOption`. An inherited class might be `AsianOption`. We might then further inherit `AsianPutOption` and `AsianCallOption` from `AsianOption`.

The key point is that each inherited class refines the class above it in the hierarchy to something more specific.

3.2 Coding inheritance

To indicate that a class, `PayOffCall`, is inherited from a class `PayOff`, we use the key word `public`. Thus when declaring the inherited class, we write

```
class PayOffCall : public PayOff
```

What effect does this have? `PayOffCall` inherits all the data members and methods of `PayOff`. And most importantly, the compiler will accept a `PayOffCall` object wherever it expects a `PayOff` object. Thus we can write all our code to work off `PayOff` objects but then use inherited class objects to specify the precise properties. (There are some issues, however; see Section 3.4.)

We can also add data members and methods at will. The rules of `public` inheritance say that we can access `protected` data and methods of the base class inside the methods of the inherited class but we cannot access the `private` data. It is generally recommended to use `private` data for this reason; if we did otherwise then any changes in the design of the base class might require that any inherited classes be rewritten and at the very least all inherited classes would have to be checked. By using `private` data, we encapsulate what the base class does, and allow ourselves to reimplement it in future. It is, however, safe to use `protected` methods, as these, similarly to `public` methods, define part of the object's interface. As long as the implicit (or ideally explicit) contract which expresses the method's functionality does not change, it does not matter what interior changes are made.

3.3 Virtual functions

Returning to our example of the `PayOff` class, we redefine the class to work using inheritance. Our base class is still called `PayOff` but whereas previously it did a lot, it now does almost nothing. In fact, it does one thing; it specifies an interface.

We give the code in `PayOff2.h` and `PayOff2.cpp`.

Listing 3.1 (`PayOff2.h`)

```
#ifndef PAYOFF2_H
#define PAYOFF2_H

class PayOff
{
public:
    PayOff(){};
    virtual double operator()(double Spot) const=0;
    virtual ~PayOff(){}
private:
```

```
};
class PayOffCall : public PayOff
{
public:
    PayOffCall(double Strike_);
    virtual double operator()(double Spot) const;
    virtual ~PayOffCall(){}

private:
    double Strike;

};

class PayOffPut : public PayOff
{
public:
    PayOffPut(double Strike_);
    virtual double operator()(double Spot) const;
    virtual ~PayOffPut(){}

private:
    double Strike;
};
#endif
```

Listing 3.2 (`PayOff2.cpp`)

```
#include <PayOff2.h>
#include <minmax.h>

PayOffCall::PayOffCall(double Strike_) : Strike(Strike_)
{
}
double PayOffCall::operator () (double Spot) const
{
    return max(Spot-Strike,0.0);
}
double PayOffPut::operator () (double Spot) const
{
    return max(Strike-Spot,0.0);
}
```

```
PayOffPut::PayOffPut(double Strike_) : Strike(Strike_)
{
}
```

The main changes to the pay-off class are that we have added the keyword virtual to operator() and we placed an =0, at the end of operator(). We have added in a destructor which is also virtual. We have also abolished all the data from both the constructor and the class definition.

We also have two new classes PayOffCall and PayOffPut. Each of these have been inherited from the class PayOff. These classes will now do all the work. The call pay-off 'is a' pay-off, and we will use PayOffCall instead of the pay-off class where we need a call pay-off. Similarly for the put pay-off.

The crucial point is that the main method, operator() is now *virtual*. What is a virtual function? In technical terms, it is a function whose address is bound at runtime instead of at compile time. What does that mean? In the code, where a PayOff object has been specified, for example in our simple Monte Carlo routine, the code when running will encounter an object of a class that has been inherited from PayOff. It will then decide what function to call on the basis of what type that object is. So if the object is of type PayOffCall, it calls the operator() method of PayOffCall, and if it is of type PayOffPut, it uses the method from the PayOffPut class and so on.

It's worth understanding how the compiler implements this. Rather than saying, "hmm, what type is this object? Ah, it's a call so I'll use the call pay-off function," it adds extra data to each object of the class which specifies what function to use. In fact, essentially what it stores is a function pointer. So virtual functions are really a fancy way of using function pointers. Indeed, if you run a program involving virtual function pointers through a debugger, and examine through the watch window an object from a class with virtual functions, you will find an extra data member, the virtual function table. This virtual function table is a list of addresses to be called for the virtual functions associated with the class. So if when running a program, a virtual function is encountered, the table is looked up and execution jumps into the function pointed to. Note that this operation takes a small amount of time so efficiency fanatics dislike virtual functions as they cause a small decrease in execution speed. Note also that the amount of memory per object has also increased as the object now contains a lot of extra data.*

---

* It is a curious fact that the C++ standard says nothing about how virtual functions are implemented so the effects are compiler dependent: However, modern compilers typically store one copy of the virtual table for each class, and each object contains a pointer to the relevant table.

If virtual functions are just function pointers, why bother with them? The first reason is that they are syntactically a lot simpler. The structure of our program is much clearer: if we can say this is a pay-off call object and this is a pay-off put object rather than having to say that we have a pay-off object but it contains a pointer to a function defining a call pay-off, we have a gain in clarity.

A second and important reason is that we get extra functionality. Depending on the complexity of the pay-off we may need extra data which cannot be expressed by a single number. With inheritance, we simply require the inherited class to contain all the data necessary. Thus for a double-digital pay-off we simply have two `doubles` expressing the upper and lower barriers. For a power option, we have a `double` for the strike and an `unsigned long` for the power. We could even have some complicated object stored as an extra argument. Indeed, if we wanted to do a complicated pay-off as a linear combination of call options, the extra data could be further call options whose pay-offs would be evaluated inside the `operator()` method and added together.

As well as being a virtual function, the `operator()` method has an =0 after it. This means that it is a *pure* virtual function. A pure virtual function has the property that it need not be defined in the base class and must be defined in an inherited class. Thus by putting =0 we are saying that the class is incomplete, and that certain aspects of its interface must be programmed in an inherited class.

In `SimpleMCMain3.cpp` we give an example of how to use the new pay-off class. Note that it uses `SimpleMC2.h` which only differs from `SimpleMC.h` in the replacement of `PayOff.h` with `PayOff2.h`. So we omit it.

Listing 3.3 (`SimpleMCMain3.cpp`)

```
/*
requires
        PayOff2.cpp
        Random1.cpp
        SimpleMC2.cpp
*/
#include<SimpleMC2.h>
#include<iostream>
using namespace std;

int main()
{
   double Expiry;
   double Strike;
   double Spot;
```

```
    double Vol;
    double r;
    unsigned long NumberOfPaths;

    cout << "\nEnter expiry\n";
    cin >> Expiry;

    cout << "\nEnter strike\n";
    cin >> Strike;

    cout << "\nEnter spot\n";
    cin >> Spot;

    cout << "\nEnter vol\n";
    cin >> Vol;

    cout << "\nr\n";
    cin >> r;

    cout << "\nNumber of paths\n";
    cin >> NumberOfPaths;

    PayOffCall callPayOff(Strike);
    PayOffPut putPayOff(Strike);

  double resultCall = SimpleMonteCarlo2(callPayOff,
                                        Expiry,
                                        Spot,
                                        Vol,
                                        r,
                                        NumberOfPaths);

    double resultPut = SimpleMonteCarlo2(putPayOff,
                                         Expiry,
                                         Spot,
                                         Vol,
                                         r,
                                         NumberOfPaths);

    cout <<"the prices are " << resultCall <<
                              " for the call and "
```

```
                        << resultPut <<
                          " for the put\n";

    double tmp;
    cin >> tmp;

    return 0;
}
```

3.4 Why we must pass the inherited object by reference

In our Monte Carlo routine, we have a parameter of type `const PayOff&` called `thePayOff`. This means that the parameter is being passed by reference rather than by value. The routine therefore works off the original object passed in. If we had not used the '&' it would copy the object: it would be passed by value not by reference. Suppose we change the parameter to type `PayOff` with or without the `const`, what happens? The code will not compile.

Why not? When the argument `Payoff` is encountered, the compiler refuses to create an object of type `PayOff` because it has a pure virtual method. The method `operator()` has not been defined, and the rules of C++ say that you cannot create an object of a type with a pure virtual method.

How could we get round this? Suppose we decide to make `operator()` an ordinary virtual function by defining it. For example, we could just make it return 0 and still override in the inherited clases as before. The code with the '&' would compile and run as before, giving the same results. The code without the '&' would now compile and run. However, the price of every option would be zero, which is certainly not what we want.

This happens because when the compiler encounters the argument of type `PayOff`, the input parameter is copied into a variable of type `PayOff`. This occurs because the compiler will call the copy constructor of `PayOff`. Like all copy constructors, it takes in an object of type `const PayOff&`. As the compiler happily accepts references to inherited objects instead of references to base-class objects this is legal and the compiler does not complain. However, the copy constructor of `PayOff` is not interested in all the extra data and information contained in the inherited object: it just looks down the data coming from the base class and copies it into the new object. As the new object is truly a base class object rather than an inherited class object pretending to be a base class object, its virtual functions will be those of the base class.

This all means that the new variable has the size and data of a `PayOff` object regardless of the type of the inputted object. The compiler therefore truncates all the additional data members which have been added, and the virtual function table is that of the base class object not the inherited class. In fact, disastrous things would occur if the new object inherited the virtual function table of the inherited object, as the virtual methods would try to access non-existent data members with possibly dangerous consequences.

Making the base class method concrete instead of pure virtual was therefore a mistake, and, in fact, the compiler's rejection of the argument without the '&' was saving us from a dangerous error.

It's worth thinking a little about what actually happens when the object is passed by reference. All that happens is that the function is passed the address of the object in memory, no copying occurs and the object's state is precisely as it was outside. However, if the object's state does change inside the routine it will also change outside which may not be what we want. We therefore include the `const` to indicate that the routine cannot do anything which may change the state of the object. The function can 'look but not touch.'

3.5 Not knowing the type and virtual destruction

Generally, one cannot pass an object of one type where another type is expected; a big feature of C++ is that it enforces type security. This means that the compiler picks up many errors at compile time instead of at run time. We cannot even run our program until we have made sure that all objects are of the types expected. This reduces the number of bugs that need to be found while the program is running by enforcing some discipline.

However, the rules of inheritance say that we can always pass a reference to an object of an inherited class instead of a reference to a base class object. This means that at times we do not know the type of object. For example, inside the Monte Carlo routine, `SimpleMC2`, we appear to be using a base class object not an inherited one.

There are two ways to forget the type of the object. The first, which we have already used, is to refer to it via a reference to the base class type instead of the inherited class type. The second, related, method is via a pointer.

If we have a pointer to an inherited object we can always cast it into a base class object without any trouble. We give an example of this in `SimpleMC-Main4.cpp`.

Listing 3.4 (`SimpleMCMain4.cpp`)

```
/*

requires

        PayOff2.cpp
```

```
        Random1.cpp
        SimpleMC2.cpp
*/
#include<SimpleMC2.h>
#include<iostream>
using namespace std;

int main()
{
    double Expiry;
    double Strike;
    double Spot;
    double Vol;
    double r;
    unsigned long NumberOfPaths;

    cout << "\nEnter expiry\n";
    cin >> Expiry;

    cout << "\nEnter strike\n";
    cin >> Strike;

    cout << "\nEnter spot\n";
    cin >> Spot;

    cout << "\nEnter vol\n";
    cin >> Vol;

    cout << "\nr\n";
    cin >> r;

    cout << "\nNumber of paths\n";
    cin >> NumberOfPaths;

    unsigned long optionType;

    cout << "\nenter 0 for call, otherwise put ";
    cin >> optionType;

    PayOff* thePayOffPtr;
```

```
if (optionType== 0)
    thePayOffPtr = new PayOffCall(Strike);
else
    thePayOffPtr = new PayOffPut(Strike);

double result = SimpleMonteCarlo2(*thePayOffPtr,
                                 Expiry,
                                 Spot,
                                 Vol,
                                 r,
                                 NumberOfPaths);

cout <<"\nthe price is " << result << "\n";
double tmp;
cin >> tmp;

delete thePayOffPtr;

return 0;
}
```

The user of our unsophisticated interface enters a zero for a call option and a non-zero number for a put option. We use an `if` statement to create the pay-off object. Note the important point here is that we use `new`. Whilst we would like to simply write

```
if (optionType== 0)
  PayOffCall thePayOff(Strike);
else
  PayOffPut thePayOff(Strike);
```

this will not give us what we want. The reason is that whilst the object thePayOff will happily be created, it will be an *automatic variable*, which is one that automatically disappears whenever the *scope* is left. In this case, the scope is the dependent clause of the `if` statement, so as soon we leave the `if` statement the object no longer exists and any attempts to reference thePayOff will result in compiler errors.

On the other hand, when we use `new` we are instructing the compiler that we wish to take some memory whilst the code is running, and that memory should not be released until we explicitly say so. The object created will therefore continue to exist outside the `if` statement as desired. The operator `new` finds enough memory to put the object into and calls the constructor, placing the object at the memory that has been allocated. It returns a pointer to the object created.

Thus the code new PayOffCall(Strike) creates the PayOffCall object and it returns a pointer to the object so we can access it. The returned pointer will be of type PayOffCall*. Fortunately, we can always cast a pointer from an inherited class pointer into a base class pointer. We therefore assign payOffPtr to the result of new and we have the desired result. Note that we declare payOffPtr outside the if to ensure that it persists after the if statement is complete.

When we reach the call to the Monte Carlo routine, we can now pass in either a call or a put, depending on what the pointer points to. We dereference the pointer in order to pass in the object rather than the pointer by putting *payOffPtr.

The final thing we must do is get rid of the object. By using new, we told the compiler that it must not destroy the object nor deallocate the memory until we say so. If we never tell it to do so then the memory will never be freed up, and our memory will slowly leak away until program crashes. The way we instruct the compiler to destroy the object and deallocate the memory is to use the operator delete. When we call delete, it first calls the destructor of the object. At this point we must be careful: we have a pointer to a base object, so which destructor will it call? If the destructor is not virtual then it will call the base class destructor. If the object is of an inherited class this may cause problems as the object will not be destroyed properly. For example, if the inherited class object had an array as a data member then the memory allocated for that array will never be deallocated. In our case, the base class is abstract and there cannot be any objects of its type. This means that calling the base class destructor must be an error. For this reason, we declare the destructor virtual. The compiler then uses the virtual function table to correctly decide which destructor to call. In fact, many compilers issue a warning if you declare a method pure virtual and do not declare the destructor virtual.

In fact, when calling and executing the destructor of an inherited class, the compiler always calls the base class destructor; this ensures that all the data members of the object which arise from the base class are correctly destroyed.

In this section, we looked at one case where we did not know the type of an object and this caused us a little trouble, but it was also very useful. By treating the inherited class object as a base class object, we were able to make the same code work regardless of the precise type of the object. The important fact was that whatever the type of the object, it had to implement all the methods defined in the base class and this was enough to ensure that the code ran smoothly. One way of summarizing this situation is to say that the inherited class *implements the interface* defined by the base class.

In the next chapter, we will look at some additional problems caused by not knowing types, such as the problem of copying an object of unknown type, and examine their solutions.

3.6 Adding extra pay-offs without changing files

One of our objectives when rewriting the PayOff class was to create a class that could be extended without changing any of the existing code. In addition, we wished to be able to add extra pay-offs without having to recompile either the file containing the PayOff class or any files which included the file defining the PayOff class. In this section, we see how to do this with our class definition.

Suppose the new pay-off we wish to add is the double digital pay-off. This pay-off pays 1 if spot is between two values and 0 otherwise. We define the new pay-off class in a new file, DoubleDigital.h with the associated code in DoubleDigital.cpp.

Listing 3.5 (DoubleDigital.h)

```
#ifndef DOUBLEDIGITAL_H
#define DOUBLEDIGITAL_H
#include <Payoff2.h>

class PayOffDoubleDigital : public PayOff
{
public:
    PayOffDoubleDigital(double LowerLevel_,
                               double UpperLevel_);

    virtual double operator()(double Spot) const;
    virtual ~PayOffDoubleDigital(){}

private:
    double LowerLevel;
    double UpperLevel;
};
#endif
```

and the source code is

Listing 3.6 (DoubleDigital.cpp)

```
#include <DoubleDigital.h>
```

```
PayOffDoubleDigital::PayOffDoubleDigital(double LowerLevel_,
                                         double UpperLevel_)
                    :    LowerLevel(LowerLevel_),
                         UpperLevel(UpperLevel_)
{
}

double PayOffDoubleDigital::operator()(double Spot) const
{
    if (Spot <= LowerLevel)
        return 0;
    if (Spot >= UpperLevel)
        return 0;

    return 1;
}
```

The crucial point is that whilst we must include the file `DoubleDigital.h` in our interface we do not include it in either the Monte Carlo file, `SimpleMC2`, or the pay-off file, `PayOff2`. The changes to the interface file are minimal and we have

Listing 3.7 (`SimpleMCMain5.cpp`)

```
/*

requires DoubleDigital.cpp
         PayOff2.cpp
         Random1.cpp
         SimpleMC2.cpp
*/
#include<SimpleMC2.h>
#include<DoubleDigital.h>
#include<iostream>
using namespace std;

int main()
{
    double Expiry;
    double Low,Up;
    double Spot;
```

```
        double Vol;
        double r;
        unsigned long NumberOfPaths;

        cout << "\nEnter expiry\n";
        cin >> Expiry;

        cout << "\nEnter low barrier\n";
        cin >> Low;

        cout << "\nEnter up barrier\n";
        cin >> Up;

        cout << "\nEnter spot\n";
        cin >> Spot;

        cout << "\nEnter vol\n";
        cin >> Vol;

        cout << "\nr\n";
        cin >> r;

        cout << "\nNumber of paths\n";
        cin >> NumberOfPaths;

        PayOffDoubleDigital thePayOff(Low,Up);

        double result = SimpleMonteCarlo2(thePayOff,
                                          Expiry,
                                          Spot,
                                          Vol,
                                          r,
                                          NumberOfPaths);

        cout <<"\nthe price is " << result << "\n";
        double tmp;
        cin >> tmp;

        return 0;
    }
```

Thus adding the new pay-off only required recompilation of one file rather than many and this achieves one objective: the `PayOff` class is open for extension but closed for modification. In particular, we could have added the new pay-off even if we did not have access to the source code of the `PayOff` class.

What we would really like is to be able to add new pay-offs without changing any files. That is we would like to be able to add new pay-offs without changing the interfacing file. In the interface, the user would simply type the name of the pay-off and this would be translated into the relevant pay-off. Rather surprisingly it is possible to do this. The solution is to use a design known as the *Factory pattern*. We shall discuss how to do this in Chapter 10.

3.7 Key points

In this chapter, we looked at how inheritance could be used to implement a `PayOff` class that is closed for modification but open for extension.

- Inheritance expresses 'is a' relationships.
- A virtual function is bound at run time instead of at compile time.
- We cannot have objects from classes with pure virtual functions.
- We have to pass inherited class objects by reference if we do not wish to change the virtual functions.
- Virtual functions are implemented via a table of function pointers.
- If a class has a pure virtual functions then it should have a virtual destructor.

3.8 Exercises

Exercise 3.1 Write an inherited class that does power options, and use it to price some.

Exercise 3.2 Implement an interface in which the user inputs a string and this is turned into a pay-off class.

Exercise 3.3 In the evil boss's list of demands in Chapter 1, try to identify as many inheritance relationships as possible.

4

Bridging with a virtual constructor

4.1 The problem

We have written a simple Monte Carlo program which uses a polymorphic class PayOff to determine the pay-off of the vanilla option to be priced. If we think about the real-world objects we wish to model, a very natural object is the option itself. At the moment, we have two pieces of data, the expiry and the pay-off, it would be much more natural to have a single piece of data, the option, which encompassed both.

How would we do this? One simple solution is simply to copy all the pay-off code we have written, add an extra data member, Expiry, to the base class and an extra method, GetExpiry, to the base class and be done. However, this rather spoils the paradigm of code reuse – we would like to reuse the same code rather than cut and paste it. In addition, if we do this for each new class of options, then it will be very time consuming to add new types of pay-offs as we will have to write a new inherited class for each option type.

A preferable solution would be to define a Vanilla Option class which has as data members a PayOff object and a double to represent expiry. However, if we try this we immediately hit a snag; the class PayOff is abstract. The compiler will squeal if we attempt to have a data member for our class of type PayOff, as you cannot instantiate an object from an abstract class. The issues we encounter here are very similar to those of Section 3.4. As there, making the PayOff class non-abstract by defining operator() in the base class causes more trouble than it solves. Any attempt to make the data member equal to an inherited class member will simply result in the object being copied and truncated into a base class object which is not what we want.

What we want is for the Vanilla Option object to be able to contain an object from an unknown class. Note that as the size of the unknown class object will not be constant, we will not be able to put it directly in the Vanilla Option object as this would lead to Vanilla Option objects being of variable size. This is something that is not allowed in C++. However, it is possible to refer to extra data outside the object using pointers or references. This is after all what an array, or a list, or a string, does.

One solution is therefore to store a reference to a pay-off object instead of a pay-off object.

4.2 A first solution

In this section, we implement a Vanilla Option class using a reference to a pay-off object and then discuss what's wrong with it.

Listing 4.1 (Vanilla1.h)

```
#ifndef VANILLA_1_H
#define VANILLA_1_H

#include <PayOff2.h>

class VanillaOption
{
public:
    VanillaOption(PayOff& ThePayOff_, double Expiry_);
    double GetExpiry() const;
    double OptionPayOff(double Spot) const;

private:
    double Expiry;
    PayOff& ThePayOff;
};
#endif
```

The source file is

Listing 4.2 (Vanilla1.cpp)

```
#include <Vanilla1.h>

VanillaOption::VanillaOption(PayOff&ThePayOff_,
                   double Expiry_)
            : ThePayOff(ThePayOff_), Expiry(Expiry_)
{
}
```

```
double VanillaOption::GetExpiry() const
{
    return Expiry;
}


double VanillaOption::OptionPayOff(double Spot) const
{
    return ThePayOff(Spot);
}
```

This is all very straightforward. The only subtlety is that the class member data is of type PayOff& instead of PayOff. However, we can use a reference to a pay-off object in precisely the same way as a pay-off object so this causes no extra coding. Our class provides two methods, one giving the expiry of the option and the other stating the pay-off at expiry given spot.

We can now rewrite the Monte Carlo routine to use the VanillaOption class. We thus have

Listing 4.3 (SimpleMC3.h)

```
#ifndef SIMPLEMC3_H
#define SIMPLEMC3_H
#include <Vanilla1.h>


double SimpleMonteCarlo3(const VanillaOption& TheOption,
                         double Spot,
                         double Vol,
                         double r,
                         unsigned long NumberOfPaths);
#endif
```

and

Listing 4.4 (SimpleMC3.cpp)

```
#include<SimpleMC3.h>
#include <Random1.h>
#include <cmath>
// the basic math functions should be in namespace
// std but aren't in VCPP6
#if !defined(_MSC_VER)
using namespace std;
#endif
```

```
double SimpleMonteCarlo3(const VanillaOption& TheOption,
                         double Spot,
                         double Vol,
                         double r,
                         unsigned long NumberOfPaths)
{
    double Expiry = TheOption.GetExpiry();

    double variance = Vol*Vol*Expiry;
    double rootVariance = sqrt(variance);
    double itoCorrection = -0.5*variance;

    double movedSpot = Spot*exp(r*Expiry +itoCorrection);
    double thisSpot;
    double runningSum=0;

    for (unsigned long i=0; i < NumberOfPaths; i++)
    {
        double thisGaussian = GetOneGaussianByBoxMuller();
        thisSpot = movedSpot*exp( rootVariance*thisGaussian);
        double thisPayOff = TheOption.OptionPayOff(thisSpot);
        runningSum += thisPayOff;
    }

    double mean = runningSum / NumberOfPaths;
    mean *= exp(-r*Expiry);
    return mean;
}
/*
```

Our new routine is very similar to the old one, the difference being that we pass in an option object instead of pay-off and an expiry separately. We obtain the expiry from the option via GetExpiry().

Our main routine is then

Listing 4.5 (VanillaMain1.cpp)

```
/*
requires DoubleDigital.cpp
         PayOff2.cpp
         Random1.cpp
         SimpleMC3.cpp
```

```
          Vanilla1.cpp
*/

#include<SimpleMC3.h>
#include<DoubleDigital.h>
#include<iostream>
using namespace std;
#include<Vanilla1.h>

int main()
{
    double Expiry;
    double Low,Up;
    double Spot;
    double Vol;
    double r;
    unsigned long NumberOfPaths;

    cout << "\nEnter expiry\n";
    cin >> Expiry;

    cout << "\nEnter low barrier\n";
    cin >> Low;

    cout << "\nEnter up barrier\n";
    cin >> Up;

    cout << "\nEnter spot\n";
    cin >> Spot;

    cout << "\nEnter vol\n";
    cin >> Vol;

    cout << "\nr\n";
    cin >> r;

    cout << "\nNumber of paths\n";
    cin >> NumberOfPaths;

    PayOffDoubleDigital thePayOff(Low,Up);
```

```
VanillaOption theOption(thePayOff, Expiry);

double result = SimpleMonteCarlo3(theOption,
                                  Spot,
                                  Vol,
                                  r,
                                  NumberOfPaths);

cout <<"\nthe price is " << result << "\n";
double tmp;
cin >> tmp;
```

```
return 0;
}
```

The main change is that now we first pass a `PayOff` object and the expiry time into a `VanillaOption` object and that is then passed into the Monte Carlo.

This program will compile and run but I do not like it. Why not? The `VanillaOption` class stores a reference to a `PayOff` object which was defined outside the class. This means that if we change that object then the pay-off of the vanilla option will change. The vanilla option will not exist as independent object in its own right but will instead always be dependent on the `PayOff` object constructed outside the class. This is a recipe for trouble. The user of the `VanillaOption` will not expect changes to the `PayOff` object to have such an effect. In addition, if the `PayOff` object had been created using `new` as we did in the last chapter then it might be `deleted` before the option ceased to exist which would result in the vanilla option calling methods of a non-existent object which is bound to cause crashes.

Similarly, if the option was created using `new` then it would likely exist long after the `PayOff` object had ceased to exist, and we would get similarly dangerous behaviour.

4.3 Virtual construction

What do we really want to do? We want the vanilla option to store its own copy of the pay-off. However, we do not want the vanilla option to know the type of the pay-off object nor anything about any of its inherited classes for all the reasons we discussed in the last chapter. Our solution there was to use virtual functions: how can we use them here? Well the object knows its own type so it can certainly make a copy of itself. Thus we define a virtual method of the base class which causes the object to create a copy of itself and return a pointer to the copy.

Such a method is called a *virtual copy constructor*. The method is generally given the name `clone`. Thus if we want the `PayOff` class to be virtually copyable, we add a pure virtual method to the base class by

```
virtual PayOff* clone() const=0;
```

and define it in each inherited class. For example,

```
PayOff* PayOffCall::clone() const
{
    return new PayOffCall(*this);
}
```

Note that the statement `new PayOffCall(*this);` says create a copy of the object for which the clone method has been called, as the `this` pointer always points to the object being called. The call to `clone` is then a call to the copy constructor of `PayOffCall` which returns a copy of the original `PayOffCall`, and because the operator `new` has been used, we can be sure that the object will continue to exist.

Note that we have made the return type of clone `PayOff*` which is a pointer to the base class object. Strictly speaking according to the standard, we should be able to change the return type to `PayOffCall*`, as it is permissible according to the standard to change the return type of a virtual function from a base class pointer to a pointer to the inherited class. As inherited class pointers can always be converted into base class pointers, this does not cause problems. However many compilers will not compile this syntax, as this is an exception to the general rule that you cannot override the return type of a virtual function.

We give the revised `PayOff` class in `PayOff3.h`. We now have

Listing 4.6 (`PayOff3.h`)

```
#ifndef PAYOFF3_H
#define PAYOFF3_H

class PayOff
{
public:
    PayOff(){};

    virtual double operator()(double Spot) const=0;
    virtual ~PayOff(){}
    virtual PayOff* clone() const=0;
```

```
private:
};

class PayOffCall : public PayOff
{
public:
    PayOffCall(double Strike_);

    virtual double operator()(double Spot) const;
    virtual ~PayOffCall(){}
    virtual PayOff* clone() const;

private:
    double Strike;
};

class PayOffPut : public PayOff
{
public:
    PayOffPut(double Strike_);

    virtual double operator()(double Spot) const;
    virtual ~PayOffPut(){}
    virtual PayOff* clone() const;

private:
    double Strike;
};
#endif
```

and `PayOff3.cpp`

Listing 4.7 (`PayOff3.cpp`)

```
#include <PayOff3.h>
#include <minmax.h>

PayOffCall::PayOffCall(double Strike_) : Strike(Strike_)
{
}
```

```
double PayOffCall::operator () (double Spot) const
{
    return max(Spot-Strike,0.0);
}


PayOff* PayOffCall::clone() const
{
    return new PayOffCall(*this);
}


double PayOffPut::operator () (double Spot) const
{
    return max(Strike-Spot,0.0);
}


PayOffPut::PayOffPut(double Strike_) : Strike(Strike_)
{
}


PayOff* PayOffPut::clone() const
{
    return new PayOffPut(*this);
}
```

We similarly have to change the contents of DoubleDigital to reflect the extra
method in the base class. We do this in DoubleDigital2.h and in
DoubleDigital2.cpp, which we do not reproduce here.

Using PayOff3.h, we can now make copies of PayOff objects of unknown type
and reimplement the VanillaOption class accordingly.

Listing 4.8 (Vanilla2.h)

```
#ifndef VANILLA_2_H
#define VANILLA_2_H


#include <PayOff3.h>


class VanillaOption
{
public:
    VanillaOption(const PayOff& ThePayOff_, double Expiry_);
```

```
    VanillaOption(const VanillaOption& original);
    VanillaOption& operator=(const VanillaOption& original);
    ~VanillaOption();

    double GetExpiry() const;
    double OptionPayOff(double Spot) const;

private:
    double Expiry;
    PayOff* ThePayOffPtr;
};
#endif
```

and the source file is

Listing 4.9 (Vanilla2.cpp)

```
#include <Vanilla2.h>

VanillaOption::VanillaOption(const PayOff&
                                ThePayOff_, double Expiry_)
                        :   Expiry(Expiry_)
{
    ThePayOffPtr = ThePayOff_.clone();
}

double VanillaOption::GetExpiry() const
{
    return Expiry;
}

double VanillaOption::OptionPayOff(double Spot) const
{
    return (*ThePayOffPtr)(Spot);
}

VanillaOption::VanillaOption(const VanillaOption& original)
{
    Expiry = original.Expiry;
    ThePayOffPtr = original.ThePayOffPtr->clone();
}
```

```
VanillaOption& VanillaOption::
                operator=(const VanillaOption& original)
{
    if (this != &original)
    {
        Expiry = original.Expiry;
        delete ThePayOffPtr;
        ThePayOffPtr = original.ThePayOffPtr->clone();
    }
    return *this;
}


VanillaOption::~VanillaOption()
{
    delete ThePayOffPtr;
}
```

We modify SimpleMC3 to get SimpleMC4 by changing the name of the include file to be Vanilla2.h and doing nothing else.

Listing 4.10 (SimpleMC4.h)

```
#ifndef SIMPLEMC4_H
#define SIMPLEMC4_H

#include <Vanilla2.h>

double SimpleMonteCarlo3(const VanillaOption& TheOption,
                        double Spot,
                        double Vol,
                        double r,
                        unsigned long NumberOfPaths);

#endif
```

Listing 4.11 (SimpleMC4.cpp)

```
#include <SimpleMC4.h>
#include <Random1.h>
#include <cmath>

// the basic math functions should be in namespace std
// but aren't in VCPP6
```

```cpp
#if !defined(_MSC_VER)
using namespace std;
#endif
double SimpleMonteCarlo3(const VanillaOption& TheOption,
                        double Spot,
                        double Vol,
                        double r,
                        unsigned long NumberOfPaths)
{

    double Expiry = TheOption.GetExpiry();

    double variance = Vol*Vol*Expiry;
    double rootVariance = sqrt(variance);
    double itoCorrection = -0.5*variance;

    double movedSpot = Spot*exp(r*Expiry +itoCorrection);
    double thisSpot;
    double runningSum=0;

    for (unsigned long i=0; i < NumberOfPaths; i++)
    {
        double thisGaussian = GetOneGaussianByBoxMuller();
        thisSpot = movedSpot*exp( rootVariance*thisGaussian);
        double thisPayOff = TheOption.OptionPayOff(thisSpot);
        runningSum += thisPayOff;
    }

    double mean = runningSum / NumberOfPaths;
    mean *= exp(-r*Expiry);
    return mean;
}
/*
```

Our main program is now VanillaMain2.cpp.

Listing 4.12 (VanillaMain2.cpp)

```cpp
/*
requires PayOff3.cpp,
        Random1.cpp,
        SimpleMC4.cpp
        Vanilla2.cpp
*/
```

```cpp
#include<SimpleMC4.h>
#include<iostream>
using namespace std;
#include<Vanilla2.h>

int main()
{
    double Expiry;
    double Strike;
    double Spot;
    double Vol;
    double r;
    unsigned long NumberOfPaths;

    cout << "\nEnter expiry\n";
    cin >> Expiry;

    cout << "\nStrike\n";
    cin >> Strike;

    cout << "\nEnter spot\n";
    cin >> Spot;

    cout << "\nEnter vol\n";
    cin >> Vol;

    cout << "\nr\n";
    cin >> r;

    cout << "\nNumber of paths\n";
    cin >> NumberOfPaths;

    PayOffCall thePayOff(Strike);

    VanillaOption theOption(thePayOff, Expiry);

    double result = SimpleMonteCarlo3(theOption,
                                      Spot,
                                      Vol,
                                      r,
                                      NumberOfPaths);
```

```
    cout <<"\nthe call price is " << result << "\n";

    VanillaOption secondOption(theOption);

    result = SimpleMonteCarlo3(secondOption,
                               Spot,
                               Vol,
                               r,
                               NumberOfPaths);

    cout <<"\nthe call price is " << result << "\n";

    PayOffPut otherPayOff(Strike);
    VanillaOption thirdOption(otherPayOff, Expiry);
    theOption = thirdOption;

    result = SimpleMonteCarlo3(theOption,
                               Spot,
                               Vol,
                               r,
                               NumberOfPaths);

    cout <<"\nthe put price is " << result << "\n";

    double tmp;
    cin >> tmp;

return 0;
}
/*
```

The actual syntax in our `main` program has not changed much.

4.4 The rule of three

One difference between the `VanillaOption` class and all the classes we have previously defined is that we have included an assignment operator (i.e. `operator=`) and a copy constructor. The reason is that whilst the compiler by default generates these methods, the compiler's definitions are inadequate.

If we do not declare the copy constructor then the compiler will perform a *shallow copy* as opposed to a *deep copy*. A shallow copy means that the data members have simply been copied, whilst this is adequate when no memory allocation is required it will lead to disaster when it does. If we have shallow copied a Vanilla-Option then both copies of the object have the same member ThePayOff-Ptr, so any modifications to the object we have pointed-to will have the same effect in each copy. Whilst for our particular object this is not really an issue as we do not want to change it, in a more complicated class this would certainly lead to trouble.

More seriously, when the one of the two VanillaOption objects goes out of scope then its destructor will be called and the pointed-to PayOff object will be deleted. The object remaining will now be in trouble as attempts to use its pointer will lead to methods of a non-existent object being called. On top of which when the remaining object goes out of scope, its destructor will be called and there will be a second attempt to delete the pointed-to object. As the object has already been deleted, this will generally result in a crash.

Thus if we write a class involving a destructor, we will generally need a copy constructor too. In fact, we have very similar issues with the assignment operator so it is necessary to write it as well. This leads to the "rule of three," which states that if any one of destructor, assignment operator and copy constructor is needed for a class then all three are. Whilst it is possible to construct examples for which this is not true, it is a very good rule of thumb. If you feel that you will never want to copy or assign objects from a class and do not bother to write the appropriate methods then you are storing up trouble; when someone does use the copy constructor the compiler will not complain but the code will crash. One solution is to declare the copy constructor but make it private. The attempt to use it will then result in a compile-time error instead of a run-time error which would be more easily spotted and fixed. (Similarly for the assignment operator.)

We analyze the copy constructor from VanillaOption. It copies the value of Expiry from the original into the copy, naturally enough. For the pointer, we call the clone method generating a copy of the PayOff object which we store in the copy's pointer. Note that the data members of the original and its copy will not be equal, since the pointers will have different values. However, this is precisely what we want, as we do not want them to be using the original PayOff object.

Note, however, that if we decided to define a comparison operator (i.e. operator==) for the class, we would have to be careful to compare the objects pointed to rather than the pointers, or we would always get false.

Having analyzed the copy constructor, let's look at the assignment operator. The first thing to note is that its return type is VanillaOption&. The effect of this is that it is legitimate to code

```
a=b=c;
```

which is equivalent to

```
b=c;
a=b;
```

but saves a little typing. The second, odder, point is that it it also legitimate to code

```
(a=b) = c;
```

which is equivalent to

```
a=b;
a=c;
```

which is not particularly useful. If we made the return type const then the second of these examples would not compile which might be regarded as an advantage; however, for built-in types the return type has no const and it is generally better to be consistent. Note that if we had made the return type void, the first example would not compile. If we had made it VanillaOption rather than Vanilla-Option& then the two examples would both compile but give odd behaviour. The first would work but might result in extra copying of objects which would waste time. The second would end with a being equal to b, as the returned object from a=b would have nothing to do with a any more and would happily be assigned the value of c and then disappear.

The second point to note about the assignment operator is that the first thing we do is check against self-assignment. If we accidentally coded a=a, we would not want the code to crash. Whilst it might seem an unlikely occurence when objects are being accessed through pointers it is not immediately obvious whether two objects are actually the same object. By checking if the address of the assigning object is equal to the this pointer, we check against this eventuality and do nothing if it occurs as nothing is needed to be done to ensure that a is equal to a.

The rest of the assignment operator is really just a combination of destructor and copy constructor. The destructor part is needed to get rid of the PayOff which the object owned before, as otherwise it would never be deleted. The copy constructor part clones the pointed-to object so the new version of the assigned part has its own copy of the object as desired.

4.5 The bridge

We have now achieved what we desired: the vanilla option class can be happily copied and moved around just like any other class and it uses the code already written for the PayOff class so we have no unnecessary duplication. However, there is still one aspect of our implementation which is slightly annoying – we had to write special code to handle assignment, construction and destruction. Every time we want to write a class with these properties, we will have to do the same thing again and again. This is rather contrary to the paradigm of reuse. What we would really like is a pay-off class that has the nice polymorphic features that our class does, whilst taking care of its own memory management. There are a couple of approaches to solving this problem. One is to use a wrapper class that has been templatized: this is really generic programming rather than object-oriented programming, and we explore this approach in Section 5.3. The other is known as *the bridge pattern*. Suppose we take the vanilla option class and get rid of the member Expiry and the method GetExpiry. We then are left with a class that does nothing except store a pointer to an option pay-off and takes care of memory handling. This is precisely what we want.

We present such a class:

Listing 4.13 (PayOffBridge.h)

```
#ifndef PAYOFFBRIDGE_H
#define PAYOFFBRIDGE_H

#include<PayOff3.h>

class PayOffBridge
{
public:
    PayOffBridge(const PayOffBridge& original);
    PayOffBridge(const PayOff& innerPayOff);

    inline double operator()(double Spot) const;
    ~PayOffBridge();
    PayOffBridge& operator=(const PayOffBridge& original);

private:
    PayOff* ThePayOffPtr;
};

inline double PayOffBridge::operator()(double Spot) const
{
    return ThePayOffPtr->operator ()(Spot);
}

#endif
```

Listing 4.14 (PayOffBridge.cpp)

```
#include<PayOffBridge.h>

PayOffBridge::PayOffBridge(const PayOffBridge& original)
{
```

```
    ThePayOffPtr = original.ThePayOffPtr->clone();
}

PayOffBridge::PayOffBridge(const PayOff& innerPayOff)
{
    ThePayOffPtr = innerPayOff.clone();
}

PayOffBridge::~PayOffBridge()
{
    delete ThePayOffPtr;
}

PayOffBridge& PayOffBridge::operator=
        (const PayOffBridge& original)
{
    if (this != &original)
    {
        delete ThePayOffPtr;
        ThePayOffPtr = original.ThePayOffPtr->clone();
    }

    return *this;
}
```

We test the new class in `VanillaMain3.cpp`. This file is almost identical to `VanillaMain2.cpp` so we do not present it here but it is included on the accompanying CD. It includes `Vanilla3.h` instead of `Vanilla2.h` and `SimpleMC5.h` instead of `SimpleMC4.h`, the only differences being the inclusion of `PayOffBridge.h`. In `Vanilla3.h`, we code the vanilla option in the way we originally wanted to! That is, we treat the pay-off as an ordinary object requiring no special treatment.

Listing 4.15 (`Vanilla3.h`)

```
#ifndef VANILLA_3_H
#define VANILLA_3_H


#include <PayOffBridge.h>


class VanillaOption
{
```

```
public:
 VanillaOption(const PayOffBridge&
                          ThePayOff_,double Expiry);


  double OptionPayOff(double Spot) const;
  double GetExpiry() const;

private:
  double Expiry;
  PayOffBridge ThePayOff;
};
#endif
```

and

Listing 4.16 (Vanilla3.cpp)

```
#include <Vanilla3.h>

VanillaOption::VanillaOption(const PayOffBridge&
                          ThePayOff_, double Expiry_)
            :  ThePayOff(ThePayOff_), Expiry(Expiry_)
{
}

double VanillaOption::GetExpiry() const
{
    return Expiry;
}

double VanillaOption::OptionPayOff(double Spot) const
{
    return ThePayOff(Spot);
}
```

Everything in Vanilla3.h is totally straightforward as all the work has already been done for by the bridged class.

An interesting aspect of VanillaMain3.cpp is that we do not have to change the lines

```
    PayOffCall thePayOff(Strike);


    VanillaOption theOption(thePayOff, Expiry);
```

Although the VanillaOption constructor expects an argument of type PayOffBridge, it happily accepts the argument of type PayOffCall. The reason is that there is a constructor for PayOffBridge which takes in an object of type PayOff&. The compiler automatically accepts the inherited class object as a substitute for the base class object, and then silently converts it for us into the PayoffBridge object which is then passed to the VanillaOption constructor.

4.6 Beware of new

We have presented the bridge as a solution to the problem of developing an open-closed pay-off class. It is a very good solution and we will use this approach again and again. However, there is a downside which it is important to be aware of:

> new is slow.

Everytime we copy a bridged object, we are implicitly using the new command. So if our code involves a lot of passing around of bridged objects, we had best be sure that we do not copy them unnecessarily, or we will spend all our time calling the new function. One easy way to reduce the amount of copying is to always pass parameters by reference.

Why is new slow? To understand this, we first have to understand what it actually does. When we create variables normally, that is when not using new, the compiler gets them from an area of memory known as the stack. The important point is that these variables are always destroyed in reverse order from their creation (which is why the memory area is called the *stack*). Each variable as declared is added to the top of the stack, and when destroyed is removed from the top; in fact, all that happens is that the code remembers a different point as the top.

This is very easy to do but really depends rather heavily on the fact that the order of creation is the reverse of the order of destruction. Thus the variable to be destroyed is always the variable at the top of the stack.

Suppose we want to destroy variables in a different order, which is what we are really doing when we use new; remember that a variable created by new persists until the coder tells it to disappear. If we were using the stack we would have to scan down to the point where the variable was stored. We would then need to move all the variables further up the stack down a bit to cover up the gap caused by the release of memory. Clearly, this could be very time-consuming.

The compiler therefore does not use the stack for new but instead uses a different area of memory known as the *heap*. For this area of memory, the code keeps track of which pieces are in use and which pieces are not. Everytime new is called, the compiler has to find an empty piece of memory which is large enough and mark the memory as being in use. When `delete` is called, the code marks the memory as being free again.

The upshot of all this is that calling `new` involves a lot of behind the scenes work which you would probably rather not have to think about.

One solution to the time-consumption of `new` is to rewrite it. This is allowed by the standard. It is, perhaps, a little surprising that one can rewrite `new` to be more efficient than a good compiler. The reason that this is possible is that the coder has a better idea of what objects will be needed during his program's run, and possibly in what order they will be created. This allows the programmer to make extra assumptions not available to the compiler, and thus to speed things up. For example, if the programmer knows that a lot of `PayOffCall` objects will be used, he could section off an area just for `PayOffCall` objects and allocate them and deallocate them as necessary.

We shall not further explore how to rewrite `new`, but finish with an admonition that it should never be used explicitly nor implicitly within time-critical loops.

4.7 A parameters class

We know that eventually the evil boss is going to demand a variable parameters Monte Carlo routine. Let's apply the ideas of this chapter to developing a `Parameters` class which allows easy extension to variable parameters without actually putting them in. Note the crucial distinction between including variable parameters, and redesigning so that it would be easy to add them if one so desired.

What should a parameters class do? We want it to be able to store parameters such as volatility or interest rates or, in a more general set-up, jump intensity. What information will we want to get out from the class? When implementing a financial model, we never actually need the instantaneous value of parameter: it is always the integral or the integral of the square that is important.

For example, in our simple Monte Carlo model, we need the square integral of the volatility from time zero to expiry, and the integral of r over the same interval. Similarly, for jump intensity it is the integral over a time interval that is important.

Our parameters class should therefore be able to tell us the integral or integral square over any time interval, and nothing else.

What sort of differing parameters might we want? It is important in object-oriented programming to think to the future, and therefore think about not just what we want now but what we might want in the future. The simplest possible sort of parameter is a constant and our class should certainly able to encompass that. Some other obvious sorts of parameters are a polynomial, a piece-wise constant function, and an exponentially decaying function. For all of these, finding the integral and square integral over definite intervals is straightforward.

We employ the bridge design. We therefore define a class `Parameters` that handles the interaction without the outside world, and the memory handling. Its only data member is a pointer to an abstract class, `ParametersInner`, which defines the interface that the concrete classes we will eventually implement must fulfill. We also include the `clone` method, as well as `Integral` and `IntegralSquare` methods, so the wrapper class, `Parameters`, can handle the copying. We then inherit classes such as `ParametersConstant` from `ParametersInner`. As with the payoff class, additional classes can then be added in separate files, and their addition will not require any recompilation or changes in routines that use the `Parameters` class.

We give a possible implementation:

Listing 4.17 (`Parameters.h`)

```
#ifndef PARAMETERS_H
#define PARAMETERS_H

class ParametersInner
{

public:
  ParametersInner(){}

    virtual ParametersInner* clone() const=0;
    virtual double Integral(double time1,
                            double time2) const=0;
    virtual double IntegralSquare(double time1,
                                  double time2) const=0;
    virtual ~ParametersInner(){}
private:
};

class Parameters
{

public:
    Parameters(const ParametersInner& innerObject);
    Parameters(const Parameters& original);
    Parameters& operator=(const Parameters& original);
    virtual ~Parameters();
```

```
    inline double Integral(double time1, double time2) const;
    inline double IntegralSquare(double time1,
                                  double time2) const;

    double Mean(double time1, double time2) const;
    double RootMeanSquare(double time1, double time2) const;

private:
    ParametersInner* InnerObjectPtr;
};

inline double Parameters::Integral(double time1,
                                    double time2) const
{
    return InnerObjectPtr->Integral(time1,time2);
}

inline double Parameters::IntegralSquare(double time1,
                                          double time2) const
{
    return InnerObjectPtr->IntegralSquare(time1,time2);
}

class ParametersConstant : public ParametersInner
{
public:
    ParametersConstant(double constant);

    virtual ParametersInner* clone() const;
    virtual double Integral(double time1, double time2) const;
    virtual double IntegralSquare(double time1,
                                  double time2) const;

private:
    double Constant;
    double ConstantSquare;
};
#endif
```

and the source file

Listing 4.18 (Parameters.cpp)

```cpp
#include <Parameters.h>

Parameters::Parameters(const ParametersInner& innerObject)
{
    InnerObjectPtr = innerObject.clone();
}


Parameters::Parameters(const Parameters& original)
{
    InnerObjectPtr = original.InnerObjectPtr->clone();
}


Parameters& Parameters::operator=(const Parameters& original)
{
    if (this != &original)
    {
        delete InnerObjectPtr;
        InnerObjectPtr = original.InnerObjectPtr->clone();
    }
    return *this;
}


Parameters::~Parameters()
{
    delete InnerObjectPtr;
}


double Parameters::Mean(double time1, double time2) const
{
    double total = Integral(time1,time2);
    return total/(time2-time1);
}


double Parameters::RootMeanSquare(double time1,
                                  double time2) const
{
    double total = IntegralSquare(time1,time2);
    return total/(time2-time1);
```

```
}

ParametersConstant::ParametersConstant(double constant)
{
    Constant = constant;
    ConstantSquare = Constant*Constant;
}

ParametersInner* ParametersConstant::clone() const
{
    return new ParametersConstant(*this);
}

double ParametersConstant::Integral(double time1,
                                    double time2) const
{
    return (time2-time1)*Constant;
}

double ParametersConstant::IntegralSquare(double time1,
                                          double time2) const
{
    return (time2-time1)*ConstantSquare;
}
```

We have added a couple of useful, if unnecessary, methods to the Parameters class. The Mean method returns the average value of the parameter between two times, and is implemented by calling the Integral method which does all the work. The RMS method returns the root-mean-square of the parameter between two times. As it is the root-mean-square of the volatility that is appropriate to use in the Black–Scholes formula when volatility is variable, this comes in useful.

The rest of the code works in the same way as for the PayOff class. We give one inherited class, the simplest possible one, ParametersConstant, which enacts a constant parameter. As well as storing the value of the constant, we also store its square in order to minimize time spent computing the square integral. In this case, the saving is rather trivial, but the principle that time can be saved by computing values, which may be needed repeatedly, once and for all in the constructor, is worth remembering.

In `SimpleMC6`, we give a modified implementation of the simple Monte Carlo which uses the new classes. First though we need

Listing 4.19 (`SimpleMC6.h`)

```
#ifndef SIMPLEMC6_H
#define SIMPLEMC6_H


#include <Vanilla3.h>
#include <Parameters.h>


double SimpleMonteCarlo4(const VanillaOption& TheOption,
                         double Spot,
                         const Parameters& Vol,
                         const Parameters& r,
                         unsigned long NumberOfPaths);


#endif
```

Listing 4.20 (`SimpleMC6.cpp`)

```
#include<SimpleMC6.h>
#include <Random1.h>
#include <cmath>


// the basic math functions should be in namespace
// std but aren't in VCPP6
#if !defined(_MSC_VER)
using namespace std;
#endif


double SimpleMonteCarlo4(const VanillaOption& TheOption,
                         double Spot,
                         const Parameters& Vol,
                         const Parameters& r,
                         unsigned long NumberOfPaths)
{

    double Expiry = TheOption.GetExpiry();
    double variance = Vol.IntegralSquare(0,Expiry);
    double rootVariance = sqrt(variance);
```

```
      double itoCorrection = -0.5*variance;

      double movedSpot = Spot*exp(r.Integral(0,Expiry) +
                                       itoCorrection);

      double thisSpot;

      double runningSum=0;

      for (unsigned long i=0; i < NumberOfPaths; i++)
      {
            double thisGaussian = GetOneGaussianByBoxMuller();

            thisSpot = movedSpot*exp( rootVariance*thisGaussian);

            double thisPayOff = TheOption.OptionPayOff(thisSpot);

            runningSum += thisPayOff;
      }

      double mean = runningSum / NumberOfPaths;

      mean *= exp(-r.Integral(0,Expiry));

      return mean;
}
```

with the obvious corresponding changes in the header file. The main difference from the previous version is that instead of computing Vol*Vol*Expiry and r*Expiry, the IntegralSquare and Integral methods of the Parameters class are called. Note that by forcing us to go via the methods of the Parameters class, the new design makes the code more comprehensible. We know immediately from looking at it that the integral of *r* is used, as is the square-integral of the volatility.

In VanillaMain4.cpp, we adapt our main program to use the Parameters class. The main difference from the previous version is that we create a ParametersConstant object from the inputs which is then passed into the Monte Carlo routine. Note that we do not have to create the Parameters object explicitly; the conversion is done implicitly by the compiler. The relevant portion of code is

```
ParametersConstant VolParam(Vol);
ParametersConstant rParam(r);

double result = SimpleMonteCarlo4(theOption,
                                  Spot,
                                  VolParam,
                                  rParam,
                                  NumberOfPaths);
```

4.8 Key points

- Cloning gives us a method of implementing a virtual copy constructor.
- The rule of three says that if we need any one of copy constructor, destructor and assignment operator then we need all three.
- We can use a wrapper class to hide all the memory handling, allowing us to treat a polymorphic object just like any other object.
- The bridge pattern allows us to separate interface and implementation, enabling us to vary the two independently.
- The new command is slow.
- We have to be careful to ensure that self-assignment does not cause crashes.

4.9 Exercises

Exercise 4.1 Test how fast new is on your computer and compiler. Do this by using it to allocate an array of doubles, ten thousand times. See how the speed varies with array size. If you have more than one compiler see how they compare. Note that you can time things using the clock() function.

Exercise 4.2 Find out about an auto_ptr. Observe that it cannot be copied in the usual sense of copying. Show that a class with an auto_ptr data member will need a copy constructor but not a destructor.

Exercise 4.3 Implement a piecewise-constant parameters class.

5

Strategies, decoration, and statistics

5.1 Differing outputs

Our Monte Carlo routine is now easily extendible to handle any pay-off and time-dependent parameters. However, there are plenty of valid criticisms that could still be made. One thing that is definitely lacking is the absence of any indication of the simulation's convergence. We could make the routine return standard error, or a convergence table, or simply have it return the value for every single path and analyze the results elsewhere.

As we are trying to develop an object-oriented routine, we make the statistics gatherer an input. Thus the Monte Carlo routine will take in a statistics gatherer object, store the results in it and the statistics gatherer will then output the statistics as required. This technique of using an auxiliary class to decide how part of an algorithm is implemented is sometimes called the *strategy pattern*.

5.2 Designing a statistics gatherer

We want our statistics gatherer to be reusable; there are plenty of circumstances where such a routine might be useful. For example, we might have many other Monte Carlo routines such as an exotics pricer or a BGM pricer for interest-rate derivatives. Also, if we are developing a risk system, we might be more interested in the ninety-fifth percentile, or in the conditional expected shortfall, than in the mean or variance.

What should the routine do? It must have two principal methods. The first should take in data for each path. The second must output the desired statistics.

Since we do not wish to specify what sort of statistics are being gathered in advance, we proceed via an abstract base class using virtual methods, just as we did for the PayOff and Parameters classes. However, as most of the time we will not need to copy these objects we do not bother with the bridge but work with the base class by reference.

We have to decide the precise interface for our two principal methods. They will be pure virtual functions declared in the base class and defined in the concrete inherited class. Our first method, DumpOneResult, we make quite simple: it takes in a double and returns nothing. Note that it is not a const method, since by its very nature it must update the statistics stored inside the object. Note that we have not allowed the possibility of dumping more than one value per path, which could be argued to be a defect. The object will store what it needs to in order to compute the statistics desired and no more. So if statistics gatherer's job is to compute the mean, then it need only store the running sum and the number of paths. However, for a more complicated statistic we might need the value for every path to be stored.

Our second method, which will indeed return the results, requires a little more thought. We have to decide what sort of object to return the results in. Another issue is whether it should be possible to ask for statistics en route or ought we be able to call the method for returning results only once.

With regard to the form in which to return results, we opt for a vector of vectors. This will allow us to easily return a table if we so desire. Whilst this would not be a great way to implement a matrix-class if we were doing linear algebra, the return statistics are not a matrix, just a table and this is sufficient for our purposes.

We opt to allow the return statistics method to be called many times and therefore name it GetResultsSoFar. This will cost us little (possibly nothing), and will be more robust than an object that crashes if the get results method is called twice. We make it a const method as it should not change the state of the object in any substantive way: this enforces the rule that the method can be called many times.

Listing 5.1 (MCStatistics.h)

```
#ifndef STATISTICS_H
#define STATISTICS_H

#include <vector>

class StatisticsMC
{
public:
    StatisticsMC(){}

    virtual void DumpOneResult(double result)=0;
    virtual std::vector<std::vector<double> >
                    GetResultsSoFar() const=0;
```

```
    virtual StatisticsMC* clone() const=0;
    virtual ~StatisticsMC(){}

private:
};

class StatisticsMean : public StatisticsMC
{

public:
    StatisticsMean();
    virtual void DumpOneResult(double result);
    virtual std::vector<std::vector<double> >
                        GetResultsSoFar() const;
    virtual StatisticsMC* clone() const;

private:
    double RunningSum;
    unsigned long PathsDone;
};
#endif
```

Our abstract base class is StatisticsMC. It has the pure virtual functions Dump-OneResult and GetResultsSoFar. We include the clone method to allow for the possibility of virtual copy construction. We also make the destructor virtual, as any cloned objects will likely need to be destroyed via pointers to the base class which will not know their type, as usual. The base class does nothing but the important task of defining an interface.

We give a very simple inherited class StatisticsMean, which returns the mean of the simulation, just as our routine previously did. The source code is included in MCStatistics.cpp.

Listing 5.2 (MCStatistics.cpp)

```
#include<MCStatistics.h>
using namespace std;

StatisticsMean::StatisticsMean()
    :
    RunningSum(0.0), PathsDone(0UL)
{
```

```
}

void StatisticsMean::DumpOneResult(double result)
{
    PathsDone++;
    RunningSum += result;
}

vector<vector<double> >
        StatisticsMean::GetResultsSoFar() const
{

    vector<vector<double> > Results(1);

    Results[0].resize(1);
    Results[0][0] = RunningSum / PathsDone;

    return Results;
}

StatisticsMC* StatisticsMean::clone() const
{
    return new StatisticsMean(*this);
}
```

Note that whilst we write the DumpOneResult method to be efficient since it will be called in every iteration of the loop, we do not worry about efficiency for GetResultsSoFar, as it will generally be called only once per simulation.

5.3 Using the statistics gatherer

Having written a class for gathering statistics, we now need to modify our Monte Carlo routine to make use of it. We do this in the function SimpleMonteCarlo5 which is declared in SimpleMC7.h.

Listing 5.3 (SimpleMC7.h)

```
#ifndef SIMPLEMC7_H
#define SIMPLEMC7_H

#include <Vanilla3.h>
#include <Parameters.h>
```

```
#include <MCStatistics.h>

void SimpleMonteCarlo5(const VanillaOption& TheOption,
                       double Spot,
                       const Parameters& Vol,
                       const Parameters& r,
                       unsigned long NumberOfPaths,
                       StatisticsMC& gatherer);
#endif
```

Note that the StatisticsMC object is passed in by reference and is not const. This is crucial as we want the object passed in to gather the information given to it and for this information to be available after the function has returned. If we had passed by value then the object outside would not change, and all the results would disappear at the end of the function which we emphatically do not want. If the object was const, then it would not be possible to put any new data into it which would be useless. Previously, our routine returned a double; now it is void as all the data to be returned is inside the object gatherer.

We define the function in SimpleMC7.cpp:

Listing 5.4 (SimpleMC7.cpp)

```
#include <SimpleMC7.h>
#include <Random1.h>
#include <cmath>
// the basic math functions should be
// in namespace std but aren't in VCPP6
#if !defined(_MSC_VER)
using namespace std;
#endif

void SimpleMonteCarlo5(const VanillaOption& TheOption,
                       double Spot,
                       const Parameters& Vol,
                       const Parameters& r,
                       unsigned long NumberOfPaths,
                       StatisticsMC& gatherer)
{
    double Expiry = TheOption.GetExpiry();
    double variance = Vol.IntegralSquare(0,Expiry);
    double rootVariance = sqrt(variance);
```

```
    double itoCorrection = -0.5*variance;
    double movedSpot =
        Spot*exp(r.Integral(0,Expiry)+itoCorrection);
    double thisSpot;
    double discounting = exp(-r.Integral(0,Expiry));

    for (unsigned long i=0; i < NumberOfPaths; i++)
    {
        double thisGaussian = GetOneGaussianByBoxMuller();
        thisSpot = movedSpot*exp( rootVariance*thisGaussian);
        double thisPayOff = TheOption.OptionPayOff(thisSpot);
        gatherer.DumpOneResult(thisPayOff*discounting);
    }
    return;
}
```

Our routine appears simpler than `SimpleMonteCarlo4`; all the work previously spent accounting the results is now sublimated into the line on which we call `.DumpOneResult`. Of course, the code has not disappeared; it has simply moved into a different file. Thus the strategy pattern gives us a readability benefit as well as flexibility.

We illustrate how the gatherer might be called in `StatsMain1.cpp`:

Listing 5.5 (StatsMain1.cpp)

```
/*
 uses source files
    MCStatistics.cpp,
    Parameters.cpp,
    PayOff3.cpp,
    PayOffBridge.cpp,
    Random1.cpp,
    SimpleMC7.cpp,
    Vanilla3.cpp,
*/
#include<SimpleMC7.h>
#include<iostream>
using namespace std;
#include<Vanilla3.h>
#include<MCStatistics.h>
```

```
int main()
{

    double Expiry;
    double Strike;
    double Spot;
    double Vol;
    double r;
    unsigned long NumberOfPaths;

    cout << "\nEnter expiry\n";
    cin >> Expiry;

    cout << "\nStrike\n";
    cin >> Strike;

    cout << "\nEnter spot\n";
    cin >> Spot;

    cout << "\nEnter vol\n";
    cin >> Vol;

    cout << "\nr\n";
    cin >> r;

    cout << "\nNumber of paths\n";
    cin >> NumberOfPaths;

    PayOffCall thePayOff(Strike);

    VanillaOption theOption(thePayOff, Expiry);

    ParametersConstant VolParam(Vol);
    ParametersConstant rParam(r);

    StatisticsMean gatherer;

    SimpleMonteCarlo5(theOption,
                      Spot,
                      VolParam,
                      rParam,
```

```
                    NumberOfPaths,
                    gatherer);

   vector<vector<double> > results = gatherer.GetResultsSoFar();

     cout <<"\nFor the call price the results are \n";

     for (unsigned long i=0; i < results.size(); i++)
     {
          for (unsigned long j=0; j < results[i].size(); j++)
              cout << results[i][j] << " ";

          cout << "\n";
     }
     double tmp;
     cin >> tmp;

     return 0;
}
```

Our output of the results is now a bit more complicated in that we have to loop over the `vector` of `vectors` in order to display the results. Of course, for the particular class we have defined, only one number is returned so this is not strictly necessary. However, by writing it for the most general sort of return statement from a statistics gatherer, we produce more robust code. After all, the object has contracted to return a `vector` of `vectors`, and we should code in accordance with this contract, and make no extra assumptions.

5.4 Templates and wrappers

We have created a class hierarchy for gathering statistics. This hierarchy includes a virtual constructor, `clone`, so we can copy these objects without knowing their type. However, if we wish to start copying and storing the objects then we have a slight issue in that, just as with our `PayOff` class, this will have to be done manually unless we write an extra wrapper class to do it for us. In the next section, we give an example where this copying will be necessary, so we do need to provide a wrapper class.

It becomes clear at this point that we will need to write these wrapper classes repeatedly in a very similar way. We therefore present a templatized solution. The name of the base class to be wrapped is taken as an argument to the wrapper class and it can then be used for any class which provides a `clone` method.

Our `Wrapper` class provides various functionalities which are intended to make it act like a pointer to a single object but with added responsibilities. The added responsibilities are that the pointer is both responsible for and owns the object pointed to at all times. Thus if we copy the `Wrapper` object, the pointed-to object is also copied, so that each `Wrapper` object has its own copy of the pointed-to object. When the `Wrapper` object ceases to exist because of going out of scope, or being `deleted`, the pointed-to object is automatically `deleted` as well.

If we set one `Wrapper` object equal to another, then the object previously pointed to must be `deleted`, and then a copy of the new object must be created so each `Wrapper` still owns precisely one object.

In addition, it must be possible to dereference the `Wrapper` to obtain the underlying object. In other words, if you put `*mywrapper` then you should obtain the object pointed to by `mywrapper`. We do this by overloading the `operator*()` and just make it return the dereferenced inner pointer.

We also want to be able to access the methods of the inner object. Whilst this can always be done by putting `(*mywrapper).theMethod()`, it is a lot less elegant than being able to type `myWrapper->theMethod()`, which is the normal code for an ordinary pointer. We provide this functionality by overloading `operator->()`.

We provide the relevant code in `Wrapper.h`.

Listing 5.6 (`Wrapper.h`)

```
#ifndef WRAPPER_H
#define WRAPPER_H

template< class T>
class Wrapper
{
public:
    Wrapper()
    { DataPtr =0;}

    Wrapper(const T& inner)
    {
        DataPtr = inner.clone();
    }

    ~Wrapper()
    {
        if (DataPtr !=0)
```

```
        delete DataPtr;
}

Wrapper(const Wrapper<T>& original)
{
    if (original.DataPtr !=0)
        DataPtr = original.DataPtr->clone();
    else
        DataPtr=0;
}

Wrapper& operator=(const Wrapper<T>& original)
{
    if (this != &original)
    {
        if (DataPtr!=0)
            delete DataPtr;

        DataPtr = (original.DataPtr !=0) ?
            original.DataPtr->clone() : 0;
    }

    return *this;
}

T& operator*()
{
    return *DataPtr;
}

const T& operator*() const
{
    return *DataPtr;
}

const T* const operator->() const
{
    return DataPtr;
}
```

```
    T* operator->()
    {
        return DataPtr;
    }

private:
    T* DataPtr;
};
#endif
```

We start before each declaration with the command template<class T> to let the compiler know we are writing template code. The class T will be specified elsewhere. The compiler will produce one copy of the code for each different sort of T that is used. Thus if we declare Wrapper<MCStatistics> TheStatsGatherer;, the compiler will then proceed to create the code by substituting MCStatistics for T everywhere, and then compile it. This has some side effects: the first is that all the code for the Wrapper template is in the header file – there is no wrapper.cpp file. The second is that if we use the Wrapper class many times, we have to compile a lot more code than we might actually expect. Whilst this is not really an issue for this class, it could be one for a complicated class, where we might end up with rather slow compile times and a much larger than expected executable. There are some other effects and we will return to this topic in Section 9.6.

We provide the Wrapper class with a default constructor. This means that it is possible to have a Wrapper object which points to nothing. If we did not then a declaration such as

```
std::vector<Wrapper<MCStatistics> > StatisticsGatherers(10);
```

would not compile: the constructor for vector would look for the default constructor for Wrapper<MCStatistics> in order to create the ten copies specified, and not find it. Why would we want a vector of Wrappers? We saw in Section 3.4 that we can get into trouble if we try to copy inherited objects into base class objects without a wrapper class. The same reasons apply here. We cannot declare a vector of base class objects as they are abstract, and even if they were not, they would be the wrong size. We therefore have to store pointers, references or wrappers, and wrappers are the easiest option; they take care of all the memory handling for us.

Given that a Wrapper object can point to nothing, we have to be able to take this into account when writing the class's methods. We indicate that the object points to nothing by setting the pointer to zero. When carrying out copying and assignment, we then have to take care of this special case.

We provide two different versions of the dereferencing operator *, as it should be possible to dereference both const and non-const objects. As one would expect, the const version returns a const object and the non-const version does not. We have declared the two operators inline to ensure that there is no performance overhead induced by going via a wrapper.

Similarly, we declare the operator-> to have both const and non-const versions. The syntax here is a little strange in that all the operator does is return the pointer. However, there are special rules for overloading -> which ensure that any method following -> is correctly invoked for the pointer returned.

5.5 A convergence table

If we use a statistics gatherer and run the simulation it will tell us the relevant statistics for the entire simulation. However, it does not necessarily give us a feel for how well the simulation has converged. One standard method of checking the convergence is to examine the standard error of the simulation; that is measure the sample standard deviation and divide by the square root of the number of paths. If one is using low-discrepancy numbers this measure does not take account of their special properties and, in fact, it predicts the same error as for a pseudo-random simulation (see for example [10]). Which we can expect to be too large. (Else why use low-discrepancy numbers?)

One alternative method is therefore to use a convergence table. Rather than returning statistics for the entire simulation, we instead return them for every power of two to get an idea of how the numbers are varying. We could just write a class directly to return such a table for the mean, but since we might want to do this for any statistic, we do it in a reusable fashion.

Our class must contain a statistics gatherer in order to decide for which statistics to create a convergence table. On the other hand, it must implement the same interface as all the other statistics gatherers so we can plug it into the same simulations. We therefore define a class ConvergenceTable which is inherited from MCStatistics, and has a wrapper of an MCStatistics object as a data member.

The fact that the class is inherited from MCStatistics guarantees that from the outside it looks just like any other statistics-gatherer object. The difference on the inside is that we can make the data member refer to any kind of statistics gatherer we like, and so we have a convergence table for any statistic for which a statistics gatherer has been written. We give the implementation in ConvergenceTable.h and ConvergenceTable.cpp.

Listing 5.7 (`ConvergenceTable.h`)

```
#ifndef CONVERGENCE_TABLE_H
#define CONVERGENCE_TABLE_H
#include <MCStatistics.h>
#include <wrapper.h>

class ConvergenceTable : public StatisticsMC
{
public:
    ConvergenceTable(const Wrapper<StatisticsMC>& Inner_);

    virtual StatisticsMC* clone() const;
    virtual void DumpOneResult(double result);
    virtual std::vector<std::vector<double> >
                            GetResultsSoFar() const;

private:
    Wrapper<StatisticsMC> Inner;
    std::vector<std::vector<double> > ResultsSoFar;
    unsigned long StoppingPoint;
    unsigned long PathsDone;
};

#endif
```

Listing 5.8 (`ConvergenceTable.cpp`)

```
#include<ConvergenceTable.h>

ConvergenceTable::ConvergenceTable(const
                        Wrapper<StatisticsMC>& Inner_)
: Inner(Inner_)
{
    StoppingPoint=2;
    PathsDone=0;
}

StatisticsMC* ConvergenceTable::clone() const
{
    return new ConvergenceTable(*this);
}
```

```
void ConvergenceTable::DumpOneResult(double result)
{
    Inner->DumpOneResult(result);
    ++PathsDone;

    if (PathsDone == StoppingPoint)
    {
        StoppingPoint*=2;
        std::vector<std::vector<double> >
            thisResult(Inner->GetResultsSoFar());

        for (unsigned long i=0; i < thisResult.size(); i++)
        {
            thisResult[i].push_back(PathsDone);
            ResultsSoFar.push_back(thisResult[i]);
        }
    }

    return;
}

std::vector<std::vector<double> >
            ConvergenceTable::GetResultsSoFar() const
{
    std::vector<std::vector<double> > tmp(ResultsSoFar);

    if (PathsDone*2 != StoppingPoint)
    {
        std::vector<std::vector<double> >
            thisResult(Inner->GetResultsSoFar());

        for (unsigned long i=0; i < thisResult.size(); i++)
        {
            thisResult[i].push_back(PathsDone);
            tmp.push_back(thisResult[i]);
        }
    }
    return tmp;
}
```

Note that we do not write a copy constructor, destructor or assignment operator as the class itself does no dynamic memory allocation. Dynamic memory allocation does occur inside the class but it is all handled automatically by the `Wrapper` template class.

The class does not do a huge amount; every result passed in is passed to the inner class. When we reach a point where the number of paths done is a multiple of two, the inner class's `GetResults()` method is called, and the results stored with the number of paths done so far added in. When the class's own `GetResults()` methods is called, it calls the inner class's method one more time if necessary and then spits out all the stored results.

In `StatsMain2.cpp`, we illustrate how the routine might be called:

Listing 5.9

```
StatisticsMean gatherer;
ConvergenceTable gathererTwo(gatherer);

SimpleMonteCarlo5(theOption,
                  Spot,
                  VolParam,
                  rParam,
                  NumberOfPaths,
                  gathererTwo);

vector<vector<double> > results =
                          gathererTwo.GetResultsSoFar();
```

First create a `StatisticsMean` object: then pass it into a `ConvergenceTable` object, gatherTwo. Note the constructor takes a `Wrapper<MCStatistics>` object but the compiler happily does this conversion for us. We then pass the new gatherer into `SimpleMonteCarlo5` which has not required any changes. We have also not made any changes to either of the `MCStatistics` files.

5.6 Decoration

The technique of the last section is an example of a standard design pattern called the *decorator pattern*. We have added functionality to a class without changing the interface. This process is called *decoration*. The most important point is that, since the decorated class has the same interface as the undecorated class, any decoration which can be applied to the original class can also be applied to the decorated class.

We can therefore decorate as many times as we wish. It would be syntactically legal (but not useful), for example, to have a convergence table of convergence tables. We will more often wish to decorate several times but in differing manners.

How else might we want to decorate? If we have a stream of numbers defining a time series, we often want the statistics of the successive increments instead of the numbers themselves. A decorator class could therefore do this differencing and pass the difference into the inner class.

We might want more than one statistic for a given set of numbers; rather than writing one class to gather many statistics, we could write a decorator class which contains a vector of statistics gatherers and passes the gathered value to each one individually. The `GetResults()` method would then garner the results from each of the inner gatherers and collate them.

We can also apply these decoration ideas to the `Parameters` class. We could define a class that takes the linear multiple of an inner `Parameters` object for example. This class would simple multiply the integral by a given constant, and the square integral by its square.

5.7 Key points

In this chapter, we have seen that we can allow the user to specify aspects of how an algorithm works by making part of the algorithm be carried out in an inputted class. We have also examined the techniques of decoration and templatization.

- Routines can be made more flexible by using the strategy pattern.
- Making part of an algorithm be implemented by an inputted class is called the strategy pattern.
- For code that is very similar across many different classes, we can use templates to save time in rewriting.
- If we want containers of polymorphic objects, we must use wrappers or pointers.
- Decoration is the technique of adding functionality by placing a class around a class which has the same interface; i.e. the outer class is inherited from the same base class.
- A class can be decorated several times.

5.8 Exercises

Exercise 5.1 Write a statistics gathering class that computes the first four moments of a sample.

Exercise 5.2 Write a statistics gathering class that computes the value at risk of a sample.

Exercise 5.3 Write a statistics gathering class that allows the computation of several statistics via inputted classes.

Exercise 5.4 Use the strategy pattern to allow the user to specify termination conditions for the Monte Carlo, e.g., time spent or paths done.

Exercise 5.5 Write a terminator class that causes termination when either of two inner terminator classes specifies termination.

Exercise 5.6 * Write a template class that implements a reference counted wrapper. This will be similar to the wrapper class but instead of making a clone of the inner object when the wrapper is copied, an internal counter is increased and the inner object is shared. When a copy is destroyed, the inner counter is decremented. When the inner counter reaches zero, the object is destroyed. Note that both the counter and the inner object will be shared across copies of the object. (This exercise is harder than most in this book.)

6

A random numbers class

6.1 Why?

So far, we have been using the inbuilt random number generator, `rand`. In this chapter, we look at how we might implement a class to encapsulate random number generation. There are a number of reasons we might wish to do this.

`rand` is implementation dependent. The standard specifies certain properties of `rand` and gives an example of how it could be implemented but it does not actually specify the details. This has important consequences for us. The first is simply that we cannot expect any consistency across compilers. If we decide to test our code by running it on multiple platforms, we can expect to obtain differing streams of random numbers and whilst our Monte Carlo simulations should still converge to the same number, this is a lot weaker than having every single random draw matching. Thus our code becomes harder to test. A second issue is that we do not know how good the compiler's implementation is. Either we have to get hold of technical documents for every compiler we use and make sure that the implementors have done a good job, or we have to run a number of statistical tests to ensure that `rand` is up to the job. Note that for most simulations we will actually need many random draws for each path, and so it is not enough for us to check that single draws do a good job of simulating the uniform distribution; instead we need a large number of successive draws to do a good job of filling out the unit hypercube, which is much tougher.

`rand` is not predictable. A crucial aspect of running Monte Carlo simulations is that they must be reproducible. If we run the same simulation twice we want to obtain precisely the same random numbers. We can achieve this with `rand` by using the `srand` command to set the seed which will guarantee the same number stream from `rand` every time. The problem is that the seed is a global variable. This means that calling `rand` in different parts of the program will cause totally unrelated pieces of code to affect each other's operation. We therefore want to be able to insulate the random number stream used by a particular simulation from the rest of the program.

Another advantage of using a class is that we can decorate it. For example, suppose we wish to use anti-thetic sampling. We could write a decorator class that does anti-thetic sampling. This can then be combined with any random number generator we have written, and plugged into the Monte Carlo simulator, with no changes to the simulator class. If we used `rand` directly we would have to fiddle with the guts of the simulator class. Similarly, if we wish to carry out moment matching we could use a decorator class and then plug the decorated class into the simulator.

A further reason is that we might decide not to use pseudo-random (i.e. random) numbers but low-discrepancy numbers instead. Low-discrepancy numbers (sometimes called quasi-random numbers) are sequences of numbers designed to do a good job of filling out space. They are therefore anything but random. However, they have the right statistical properties to guarantee that simulations converge to the correct answer. Their space-filling properties mean they make simulations converge faster. If we are using a random number class, we could replace this class with a generator for low-discrepancy numbers without changing the interior of our code.

6.2 Design considerations

As we want the possibility of having many random number generators and we want to be able to add new ones later on without recoding, we use an abstract base class to specify an interface. Each individual generator will then be inherited from it. In order to specify the interface, we have to identify what we want from any random number class.

Generally, when working with any Monte Carlo simulation, the simulation will have a dimensionality which is the number of random draws needed to simulate one path. This number is equal to the number of variables of integration in the underlying integral which we are trying to approximate. It is generally cleaner therefore to obtain all the draws necessary for a path in one go. This has the added advantage that a random number generator can protest (i.e. throw an error) if it is being used beyond its dimensional specification. Additionally, when working with low-discrepancy numbers it is essential that the generator know the dimensionality as the generator has to be set up specifically for each choice of dimension.

This means that we need methods to set the dimensionality, and to obtain an array of uniforms of size dimensionality from the generator. We also provide a method that states the dimensionality.

For financial applications, we will want standard Gaussian draws more often than uniforms so we will want a method of obtaining them instead. In fact, we can separate out the creation of the uniforms and their conversion into Gaussians. The conversion into Gaussians can therefore be done in a generator-independent fashion and this means that it can be implemented as a method of the base class which calls the virtual method that creates the uniform draws.

What else might we want? For many applications, it is necessary to generate the same stream of random numbers twice. For example, if we wish to compute Greeks by bumping parameters, the error is much smaller if the same numbers are used twice. (See for example [11] or [13].) Or if we wish to carry out moment matching, the reuse of the same random numbers stream twice enables us to avoid storing all the numbers generated. Thus we include methods to reset the generator to its initial state, and to set the seed of the generator.

Occasionally, we wish to be sure of having a different stream of random numbers. For example, when carrying out an optimization in order to estimate an exercise strategy, we generally use one set of random numbers to optimize parameters for the strategy, and then having chosen the strategy we run a second simulation with different random numbers to estimate the price. This allows us to be sure that the optimization has not exploited the micro-structure of the random number stream. A simple way to achieve the differing streams of numbers is to make sure the generator skips a number of paths equal to the number used for the first simulation. We therefore include a method which allows us to skip paths.

Finally, we may wish to copy a random number generator for which we do not know the type. We therefore include a `clone` method to enable virtual construction.

One extra issue we have to think about is in what range a uniform should lie. The uniform distribution is generally defined to be a density function on the interval [0, 1] such that the probability that a draw X lies in an interval of length α is α. The subtlety lies in whether we allow the values 0 and 1 to be taken. Since taking either value is a probability zero event allowing or disallowing either value will not effect the statistical properties of our simulation, but they can have practical effects. For example, if we elect to convert the uniforms into Gaussians by using the inverse cumulative normal function (which we will) then the numbers 0 and 1 cause us difficulties since the inverse cumulative normal function naturally maps them to $-\infty$ and $+\infty$. To avoid these difficulties, we therefore require that our uniform variates never take these values and thus lie in the open interval $(0, 1)$. The main side effect of this choice is that if we use random generators written by others then we need to check that they satisfy the same convention, and if not, adapt them appropriately.

6.3 The base class

We specify the interface via a base class as follows, Random2.h,

Listing 6.1 (Random2.h)

```
#ifndef RANDOM2_H
#define RANDOM2_H

#include <Arrays.h>

class RandomBase
{
public:
    RandomBase(unsigned long Dimensionality);

    inline unsigned long GetDimensionality() const;

    virtual RandomBase* clone() const=0;
    virtual void GetUniforms(MJArray& variates)=0;
    virtual void Skip(unsigned long numberOfPaths)=0;
    virtual void SetSeed(unsigned long Seed) =0;
    virtual void Reset()=0;

    virtual void GetGaussians(MJArray& variates);
    virtual void ResetDimensionality(unsigned long
                                    NewDimensionality);

private:
    unsigned long Dimensionality;
};

unsigned long RandomBase::GetDimensionality() const
{
    return Dimensionality;
}
#endif
```

Whilst most of the methods of RandomBase are pure virtual, three are not. The method GetGaussians transforms uniforms obtained from the GetUniforms method into standard Gaussian distributions. It does this via an approximation to the inverse cumulative normal function due to Moro, [21]. As this method only uses one uniform to produce a Gaussian and enacts precisely the definition of the Gaussian distribution it is very robust and works under all circumstances. Nevertheless, we make the method virtual to allow the possibility that for a particular generator there is another preferred conversion method. Or even to allow the possibility that the generator provides normals which are then converted into uniforms by the `GetUniforms` method.

The `GetDimensionality` method simply returns the dimensionality of the generator and there is no need for it to be virtual.

We also have the concrete virtual function `ResetDimensionality`. As the base class stores dimensionality, it must be told when dimensionality changes: that is the purpose of this function. However, the function is virtual because generally if dimensionality changes, the random number generator will also need to know. Suppose we have overriden this virtual function in an inherited class. Calling the method thus only calls the inherited class method and the base class method is ignored; however, we still need the base class method to be called; this has to be done by the inherited class method. The syntax to do this is to prefix the method with `RandomBase::`. The compiler then ignores the virtual function table and instead knows to call the method associated to the base class.

Note that we define the interface for `GetUniforms` and `GetGaussians` via a reference to an array. The reason we do this is that we do not wish to waste time copying arrays. Also remember that arrays of dynamic size generally involve dynamic memory allocation, i.e. `new`, and therefore are quite slow to create and to destroy. We want to minimize unnecessary operations, and by passing the return values into a pre-generated array we avoid all this. The array class used here is quite simple and given in Appendix C. We assume that the array is of sufficient size. We could check that it is big enough but that could result in substantial overhead. One solution would be to check the size only if a compiler flag was set, e.g. in debug mode.

Note that one disadvantage of this approach is that we are now bound to this array class. How could we overcome that disadvantage? One solution would be to simply pass in a pointer, and write to the memory locations pointed to. However, the use of raw pointers tends to lead to code that is hard to debug, and is therefore best avoided. Another solution is to templatize so that the array class is a template argument and the code will then work with any array class which has the requisite methods. A related solution is to use iterators. An *iterator* is a generalization of a pointer and we could templatize the code to work off any iterator. We do not explore these options here but the reader should bear them in mind if he wishes to adapt the code.

The source code for the base class is quite simple as it does not do very much:

Listing 6.2 (Random2.cpp)

```cpp
#include <Random2.h>
#include <Normals.h>
#include <cstdlib>

// the basic math functions should be in namespace
// std but aren't in VCPP6
#if !defined(_MSC_VER)
using namespace std;
#endif

void RandomBase::GetGaussians(MJArray& variates)
{
    GetUniforms(variates);

    for (unsigned long i=0; i < Dimensionality; i++)
    {
        double x=variates[i];
        variates[i] = InverseCumulativeNormal(x);
    }
}

void RandomBase::ResetDimensionality(unsigned long
                                        NewDimensionality)
{
    Dimensionality = NewDimensionality;
}

RandomBase::RandomBase(unsigned long Dimensionality_)
: Dimensionality(Dimensionality_)
{
}
```

The inverse cumulative normal function is included in the file Normals and is a piece-wise rational approximation. See Appendix B.

6.4 A linear congruential generator and the adapter pattern

We now need to actually write a random number generator. A simple method of generating random numbers is a linear congruential generator. We present a generator called by Park & Miller the *minimal standard generator*. In other words, it is a generator that provides a minimum guaranteed level of statistical accuracy. We refer the reader to [28] for further discussion of this and many other random number generators.

We present the generator in two pieces. A small inner class that develops a random generator that returns one integer (i.e., `long`) every time it is called, and a larger class that turns the output into a vector of uniforms in the format desired. We present the class definition in `ParkMiller.h`.

Listing 6.3 (ParkMiller.h)

```
#ifndef PARK_MILLER_H
#define PARK_MILLER_H
#include <Random2.h>

class ParkMiller
{
public:
    ParkMiller(long Seed = 1);

    long GetOneRandomInteger();
    void SetSeed(long Seed);

    static unsigned long Max();
    static unsigned long Min();

private:
    long Seed;
};

class RandomParkMiller : public RandomBase
{
public:
    RandomParkMiller(unsigned long Dimensionality,
                     unsigned long Seed=1);

    virtual RandomBase* clone() const;
    virtual void GetUniforms(MJArray& variates);
    virtual void Skip(unsigned long numberOfPaths);
    virtual void SetSeed(unsigned long Seed);
    virtual void Reset();
```

```
    virtual void ResetDimensionality(unsigned long
                                        NewDimensionality);

private:
    ParkMiller InnerGenerator;
    unsigned long InitialSeed;
    double Reciprocal;
};
#endif
```

The inner class is quite simple. It develops a sequence of uncorrelated longs. The seed can be set either in the constructor or via a set seed method. We give two extra methods which indicate the minimum and maximum values that the generator can give out. Such information is crucial to a user who wishes to convert the output into uniforms, as they will need to subtract the minimum and divide by the maximum minus the minimum to get a number in the interval [0, 1].

The bigger class is inherited from RandomBase. It has all the methods that it requires. Its main data member is a ParkMiller generator object. It also remembers the initial seed, and the reciprocal of the maximum value plus one, to save time then turning the output of the inner generator into uniforms.

Our pattern here is an example of the *adapter* pattern. We have a random generator which works and is effective, however its interface is not what the rest of the code expects. We therefore write a class around it which adapts its interface into what we want. Whenever we use old code or import libraries, it is rare for the interfaces to fit precisely with what we have been using, and the adapter pattern is then necessary. To use the adapter pattern simply means to use an intermediary class which transforms one interface into another. It is the coding equivalent of a plug adapter.

The implementation of these classes is straightforward. The generator relies on modular arithmetic. The basic idea is that if you repeatedly multiply a number by a large number, and then take the modulus with respect to another number, then the successive remainders are effectively random. We refer the reader to [28] for discussion of the mathematics and the choice of the constants.

Listing 6.4 (ParkMiller.cpp)

```
#include <ParkMiller.h>

const long a = 16807;
const long m = 2147483647;
const long q = 127773;
const long r = 2836;
```

```
ParkMiller::ParkMiller(long Seed_ ) : Seed(Seed_)
{
    if (Seed ==0)
        Seed=1;
}

void ParkMiller::SetSeed(long Seed_)
{
  Seed=Seed_;
  if (Seed ==0)
        Seed=1;
}

unsigned long ParkMiller::Max()
{
    return m-1;
}

unsigned long ParkMiller::Min()
{
    return 1;
}

long ParkMiller::GetOneRandomInteger()
{
    long k;

    k=Seed/q;

    Seed=a*(Seed-k*q)-r*k;

    if (Seed < 0)
        Seed += m;

    return Seed;
}

RandomParkMiller::RandomParkMiller(unsigned long Dimensionality,
                                   unsigned long Seed)
:    RandomBase(Dimensionality),
    InnerGenerator(Seed),
```

```
    InitialSeed(Seed)
{
    Reciprocal = 1/(1.0+InnerGenerator.Max());
}


RandomBase* RandomParkMiller::clone() const
{
    return new RandomParkMiller(*this);
}


void RandomParkMiller::GetUniforms(MJArray& variates)
{
    for (unsigned long j=0; j < GetDimensionality(); j++)
        variates[j] =
        InnerGenerator.GetOneRandomInteger()*Reciprocal;
}


void RandomParkMiller::Skip(unsigned long numberOfPaths)
{
    MJArray tmp(GetDimensionality());
    for (unsigned long j=0; j < numberOfPaths; j++)
        GetUniforms(tmp);
}


void RandomParkMiller::SetSeed(unsigned long Seed)
{
    InitialSeed = Seed;
    InnerGenerator.SetSeed(Seed);
}


void RandomParkMiller::Reset()
{
    InnerGenerator.SetSeed(InitialSeed);
}


void RandomParkMiller::ResetDimensionality(unsigned long
                                          NewDimensionality)
{
    RandomBase::ResetDimensionality(NewDimensionality);
    InnerGenerator.SetSeed(InitialSeed);
}
```

Note that we check whether the seed is zero. If it is we change it to 1. The reason is that a zero seed yields a chain of zeros. Note the advantage of a class-based implementation here. The seed is only inputted in the constructor and the set seed method, which are called only rarely, so we can put in extra tests to make sure the seed is correct with no real overhead. If the seed had to be checked every time the random number generator was called, then the overhead would be substantial indeed.

The implementation of the adapter class is quite straightforward. Note that we divide the outputs of the inner class by the maximum plus 1, and so ensure that we obtain random numbers on the open interval (0, 1) rather than the closed one; this means that we will have no trouble with the inverse cumulative normal function.

6.5 Anti-thetic sampling via decoration

A standard method of improving the convergence of Monte Carlo simulations is anti-thetic sampling. The idea is very simple, if a X is a draw from a standard Gaussian distribution so is $-X$. This means that if we draw a vector $(X_1, \ldots, X_n)$ for one path then instead of drawing a new vector for the next path we simply use $(-X_1, \ldots, -X_n)$. This method guarantees that, for any even number of paths drawn, all the odd moments of the sample of Gaussian variates drawn are zero, and in particular the mean is correct. This generally, but not always, causes simulations to converge faster. See [11] for discussion of the pros and cons of anti-thetic sampling.

We wish to implement anti-thetic sampling in such a way that it can be used with any random number generator and with any Monte Carlo simulation in such a way that we only have to implement it once. The natural way to do this is the decorator pattern. The decoration can be applied to any generator so it fulfills the first criterion, and the fact that the interface is unchanged means that we can plug the decorated class into any socket which the original class fitted. We implement such a decorator class in AntiThetic.h and AntiThetic.cpp.

Listing 6.5 (AntiThetic.h)

```
#ifndef ANTITHETIC_H
#define ANTITHETIC_H

#include <Random2.h>
#include <wrapper.h>

class AntiThetic : public RandomBase
{
```

```
public:

    AntiThetic(const Wrapper<RandomBase>& innerGenerator );

    virtual RandomBase* clone() const;

    virtual void GetUniforms(MJArray& variates);

    virtual void Skip(unsigned long numberOfPaths);

    virtual void SetSeed(unsigned long Seed);

    virtual void ResetDimensionality(unsigned long
                                        NewDimensionality);

    virtual void Reset();
private:
    Wrapper<RandomBase> InnerGenerator;

    bool OddEven;

    MJArray NextVariates;
};
#endif
```

The decorator class is quite simple. It has an array as a data member to store the last vector drawn, and a boolean to indicate whether the next draw should be drawn from the inner generator, or be the anti-thetic of the last draw. A copy of the generator we are using is stored using the Wrapper template class and cloning, as usual. Note that we are actually taking a copy of the generator here so that the sequence of draws from the original generator will not be affected by drawing from the anti-thetic generator.

Listing 6.6 (AntiThetic.cpp)

```
#include <AntiThetic.h>

AntiThetic::AntiThetic(const Wrapper<RandomBase>&
                                        innerGenerator )
        : RandomBase(*innerGenerator),
          InnerGenerator(innerGenerator)
{
```

```cpp
    InnerGenerator->Reset();
    OddEven =true;
    NextVariates.resize(GetDimensionality());
}


RandomBase* AntiThetic::clone() const
{
    return new AntiThetic(*this);
}


void AntiThetic::GetUniforms(MJArray& variates)
{
    if (OddEven)
    {
        InnerGenerator->GetUniforms(variates);

        for (unsigned long i =0; i < GetDimensionality(); i++)
            NextVariates[i] = 1.0-variates[i];

        OddEven = false;
    }
    else
    {
        variates = NextVariates;

        OddEven = true;
    }
}


void AntiThetic::SetSeed(unsigned long Seed)
{
    InnerGenerator->SetSeed(Seed);
    OddEven = true;
}


void AntiThetic::Skip(unsigned long numberOfPaths)
{
    if (numberOfPaths ==0)
        return;

    if (OddEven)
```

```
    {
        OddEven = false;
        numberOfPaths--;
    }

    InnerGenerator->Skip(numberOfPaths  / 2);

    if (numberOfPaths % 2)
    {
        MJArray tmp(GetDimensionality());

        GetUniforms(tmp);

    }
}

void AntiThetic::ResetDimensionality(unsigned long
                                                NewDimensionality)

{
    RandomBase::ResetDimensionality(NewDimensionality);

    NextVariates.resize(NewDimensionality);

    InnerGenerator->ResetDimensionality(NewDimensionality);
}

void AntiThetic::Reset()
{
    InnerGenerator->Reset();
    OddEven =true;
}
```

The implementation of the class is quite straightforward. Most of the methods consist of simply forwarding the request to the inner class, together with book-keeping for odd and even paths. The main `GetUniforms` method, gets uniforms from the inner generator for the odd draws, stores the results, X_j, and returns $(1 - X_1, \ldots, 1 - X_n)$ for the even draws. Note that

$$N^{-1}(1 - x) = -N^{-1}(x), \tag{6.1}$$

so this will yield the negative of the Gaussian variates if the `GetGaussians` method is used, as we wanted.

Note the syntax for initialization in the constructor. We have RandomBase
(*innerGenerator). As innerGenerator is a wrapped pointer, * gives us the
value of the inner object which is a member of some inherited class. However,
we can always treat any inherited class object as a base class object so the call
to RandomBase invokes the base class copy constructor, copying the base class
data in innerGenerator, and thus ensuring that the new object has the correct
dimensionality stored.

6.6 Using the random number generator class

Now that we have a random number generator class, we need to adapt our Monte
Carlo code to work with it. We give an adapted vanilla option pricer in
SimpleMC8.h and SimpleMC8.cpp. The header file declares the new func-
tion.

Listing 6.7 (SimpleMC8.h)

```
#ifndef SIMPLEMC8_H
#define SIMPLEMC8_H


#include <Vanilla3.h>
#include <Parameters.h>
#include <Random2.h>
#include <MCStatistics.h>


void SimpleMonteCarlo6(const VanillaOption& TheOption,
                       double Spot,
                       const Parameters& Vol,
                       const Parameters& r,
                       unsigned long NumberOfPaths,
                       StatisticsMC& gatherer,
                       RandomBase& generator);
#endif
```

We have chosen to take the random number generator in as a non-const reference.
It cannot be a const reference as the act of drawing a random number changes the
generator and is therefore implemented by a non-const method. The effect of this
is that any random numbers drawn inside the function will not be produced outside
the function, but instead the generator will continue where the function left off.
If we wanted the generator to be totally unaffected by what happened inside the
function, we would change the function to take in the object by value instead. Or
alternatively, we could copy the object and pass in the copy to the function, which

would have the same net effect. As usual, we use a reference to the base class in
order to allow the caller to decide how to implement the generator.

The implementation is as follows:

Listing 6.8 (`SimpleMC8.cpp`)

```cpp
#include<SimpleMC8.h>
#include <cmath>
#include <Arrays.h>

// the basic math functions should be in
// namespace std but aren't in VCPP6
#if !defined(_MSC_VER)
using namespace std;
#endif

void SimpleMonteCarlo6(const VanillaOption& TheOption,
                       double Spot,
                       const Parameters& Vol,
                       const Parameters& r,
                       unsigned long NumberOfPaths,
                       StatisticsMC& gatherer,
                       RandomBase& generator)
{
    generator.ResetDimensionality(1);

    double Expiry = TheOption.GetExpiry();
    double variance = Vol.IntegralSquare(0,Expiry);
    double rootVariance = sqrt(variance);
    double itoCorrection = -0.5*variance;
    double movedSpot = Spot*exp(r.Integral(0,Expiry)
                              + itoCorrection);

    double thisSpot;
    double discounting = exp(-r.Integral(0,Expiry));

    MJArray VariateArray(1);

    for (unsigned long i=0; i < NumberOfPaths; i++)
    {
```

```
    generator.GetGaussians(VariateArray);
    thisSpot = movedSpot*exp(rootVariance*VariateArray[0]);
    double thisPayOff = TheOption.OptionPayOff(thisSpot);
    gatherer.DumpOneResult(thisPayOff*discounting);
}


    return;
}
```

We only comment on the new aspects of the routine. We first reset the generator's
dimensionality to 1 as pricing a vanilla option is a one-dimensional integral – we
just need the location of the final value of spot.

   We set up the array in which to store the variate before we set up the main
loop, once and for all. This avoids any difficulties with speed in the allocation of
dynamically sized arrays. The GetGaussians method of the generator is used
to write the variates (in this case just one variate, of course) into the array. This
variate is then used as before to compute the final value of spot.

   We give an example of using this routine with anti-thetic sampling in Random-
Main3.cpp.

Listing 6.9 (RandomMain3.cpp)

```
/*
    uses source files
    AntiThetic.cpp
    Arrays.cpp,
    ConvergenceTable.cpp,
    MCStatistics.cpp
    Normals.cpp
    Parameters.cpp,
    ParkMiller.cpp
    PayOff3.cpp,
    PayOffBridge.cpp,
    Random2.cpp,
    SimpleMC8.cpp
    Vanilla3.cpp,
*/
#include<SimpleMC8.h>
#include<ParkMiller.h>
#include<iostream>
```

```
using namespace std;
#include<Vanilla3.h>
#include<MCStatistics.h>
#include<ConvergenceTable.h>
#include<AntiThetic.h>

int main()
{

    double Expiry;
    double Strike;
    double Spot;
    double Vol;
    double r;
    unsigned long NumberOfPaths;

    cout << "\nEnter expiry\n";
    cin >> Expiry;

    cout << "\nStrike\n";
    cin >> Strike;

    cout << "\nEnter spot\n";
    cin >> Spot;

    cout << "\nEnter vol\n";
    cin >> Vol;

    cout << "\nr\n";
    cin >> r;

    cout << "\nNumber of paths\n";
    cin >> NumberOfPaths;

    PayOffCall thePayOff(Strike);

    VanillaOption theOption(thePayOff, Expiry);

    ParametersConstant VolParam(Vol);
    ParametersConstant rParam(r);
```

```
StatisticsMean gatherer;
ConvergenceTable gathererTwo(gatherer);

RandomParkMiller generator(1);

AntiThetic GenTwo(generator);

SimpleMonteCarlo6(theOption,
                                Spot,
                                VolParam,
                                rParam,
                                NumberOfPaths,
                                gathererTwo,
                                GenTwo);

vector<vector<double> > results =
     gathererTwo.GetResultsSoFar();

cout <<"\nFor the call price the results are \n";
for (unsigned long i=0; i < results.size(); i++)
{
     for (unsigned long j=0; j < results[i].size(); j++)
          cout << results[i][j] << " ";

     cout << "\n";
}
double tmp;
cin >> tmp;

   return 0;
}
```

We create a Park–Miller random number generator object and then wrap it with an anti-thetic decorator. This decorated object is then passed into the new Monte Carlo routine. As usual, the routine is not aware of the fact that the passed-in object has been decorated but simply uses it in the same way as any other random number generator.

The big difference between our new program and the old ones is that the results are now compiler-independent. The numbers returned are now precisely the same under Borland 5.5, Visual C++ 6.0 and MingW 2.95, since we have removed the dependency on the inbuilt `rand()` function which previously made our results compiler-dependent. This gives us an extra robustness test; if our results are not now compiler-independent we should be worried and find out why!

6.7 Key points

In this chapter, we developed a random number generator class and saw how anti-thetic sampling could be implemented via decoration.

- `rand` is implementation-dependent.
- Results from `rand()` are not easily reproducible.
- We have to be sure that a random generator is capable of the dimensionality necessary for a simulation.
- Using a random number class allows us to use decoration.
- The inverse cumulative normal function is the most robust way to turn uniform variates from the open interval, (0, 1), into Gaussian variates.
- Using a random number class makes it easier to plug in low-discrepancy numbers.
- Anti-thetic sampling can be implemented via decoration.

6.8 Exercises

Exercise 6.1 For various cases compare convergence of Monte Carlo simulations with and without anti-thetic sampling.

Exercise 6.2 Obtain another random number generator and fit it into the class hierarchy given here. (See [28] or `www.boost.org` for other generators.)

Exercise 6.3 Code up a low-discrepancy number generator and integrate it into the classes here. (See [28] or [11].)

7

An exotics engine and the template pattern

7.1 Introduction

We have now developed quite a few sets of components: random number generators, parameters classes, pay-off classes, statistics gatherers and a wrapper template. Having developed all these components with the objective of reusability, we examine in this chapter how to put them together to price path-dependent exotic options. Our objective is to develop a flexible Monte Carlo pricer for exotic options which pay off at some future date according to the value of spot on a finite number of dates. We will work within a deterministic interest rate world, and assume the Black–Scholes model of stock price evolution.

We assume that our derivative is discrete, i.e. that it depends upon the value of spot on a discrete set of times. Thus our derivative is associated to a set of times, $t_1, t_2, \ldots, t_n$, and pays at some time T a function $f(S_{t_1}, \ldots, S_{t_n})$ of the value of spot at those times. For example, a one-year Asian call option struck at K with monthly resets would pay

$$\left(\frac{1}{12} \sum_{j=1}^{12} S_{t_j} - K \right)_+ ,$$

where $t_j = j/12$, at time 1.

More generally, the derivative could possibly pay cash-flows at more than one time. For example, a discrete barrier knock-out option could pay an ordinary vanilla pay-off at the time of expiry, and a rebate at the time of knock-out otherwise.

We do not consider American or Bermudan options here as the techniques involved are quite different. Note, however, that once an exercise strategy has been chosen the option is really just a path-dependent derivative and so the option can be evaluated by these techniques for any given fixed exercise strategy.

7.2 Identifying components

Before designing our engine let's identify what it will have to do. For each path, we generate a discounted pay-off which is then averaged over all paths to obtain a price. To generate a pay-off for a path, we have to know the path of stock prices at the relevant times, plug this path into the pay-off function of the option to obtain the cash-flows, and then discount these cash-flows back to the start to obtain the price for that path.

We can therefore identify four distinct actions:

 (i) the generation of the stock price path;
 (ii) the generation of cash-flows given a stock price path;
(iii) the discounting and summing of cash-flows for a given path;
(iv) the averaging of the prices over all the paths.

We already have a suitable component for the last of these actions: the statistics gatherer. We will just plug in the class we have already written at the appropriate point. For the second action, we are purely using the definition of the derivative to determine what its pay-off is, given a set of stock prices. An obvious component for our model is therefore a path-dependent exotic option class which will encapsulate the information which would be written in the *term-sheet* for the option.

Note that by defining the concept we model by the term-sheet, we make it clear that the class will not involve interest rates nor knowledge of volatility nor any aspect of the stock price process. A consequence of this is that the option class can only ever say what cash-flows occur and when; it cannot say anything about their discounted values because that would require knowledge of interest rates. Note the general point here that, by defining the concept in real-world terms, it becomes very easy to decide what should and should not be contained in the class. Another consequence is that the component is more easily reusable; if we decide to do jump-diffusion pricing or stochastic interest rates, this class will be reusable without modification, and that would not be the case if we had included information about interest rates or volatility.

There is more than one way to handle the two remaining tasks: path generation and cash-flow accounting. The latter will be the same for any deterministic interest-rate model and it is therefore natural to include it as part of our main engine class. We can require path generation to be an input to our main class, and therefore define it in terms of its own class hierarchy, or via a virtual method of the base class. A third option, which we do not explore, would be to make it a template parameter, which would avoid the small overhead of a virtual function call.

The option we will pursue here is to make path generation a virtual method of the base class. This is an example of the *template* design pattern, which should not be confused with templatized code. The idea here is that the base class sets up a structure with methods that control everything – in this case it would be methods to run the simulation and account for each path – though the main work of actually generating the path is not defined in the base class, but is instead called via a pure virtual function. This pure virtual function must therefore be defined in an inherited class. Thus as in templatized code, the code is set up more to define a structure than to do the real work which is coded elsewhere.

7.3 Communication between the components

Having identified what our components will be, we still need to assess what information has to be passed between them, and we need to decide how to carry out the communication.

The product will take in a vector of spot values for its relevant times and spit out the cash-flows generated. A couple of immediate consequences of this are that there has to be a mechanism for the product to tell the path generator for which times it needs the value of spot, and that we need to decide how to define a cash-flow object.

To deal with the first of these, we include a method `GetLookAtTimes`; this passes back an array of times that are relevant to the pay-off function of the product. For the definition of cash-flow objects, there are a couple of options. The first obvious approach is to simply make a cash-flow a pair of `doubles` which define the amount and the time of the cash-flow. This approach has the disadvantage that if we have a complicated term structure of interest rates, the action of computing the discount factor for the cash-flow time may be slow, and with the product being allowed to pass back an arbitrary time, this discounting will have to be done on every path. Whilst one could cache already-computed times, there is then still the problem that searching the cache for the already-computed times will take time.

In practice, many products can only pay off at one time. This means that it would be better to pre-compute the discount factor for that time. However, we would still need to know in advance what that time is. We therefore require our product to have another method, `PossibleCashFlowTimes`, which returns an array defining the possible times. As the engine will know all the possible times in advance we can return a cash-flow as a pair: an index and an amount. The index is now an `unsigned long` instead of a `double`. The engine will now precompute all the discount factors and then simply use the index to look up an array to get the discounting for any given cash-flow.

We still have to decide the syntax for the main method `CashFlows`. The method takes in an array defining spot values, and returns cash-flows. As we allow the possibility of more than one cash-flow, we must use a container to pass them back. We use the STL vector class. Whilst it would be tempting to make the return type of the class `vector<CashFlow>`, this would have timing disadvantages. We would have to create a new vector every time the method was called, and this could be time consuming because any dynamically sized container class must involve memory allocation. We therefore take an argument of type vector<CashFlow>& into which we write the cash-flows.

We still have the issue that the vector will need to be of the correct size. One solution is for the method to resize it as necessary but this could be time consuming. First, resizing can involve memory allocation though this is not a huge issue since the memory allocated for an STL vector never shrinks so if the same vector is used every time it will rapidly grab enough memory and then will need no more. Second, some implementations of the STL explicitly destroy all the objects in the vector during a resize, which means that every resize involves looping, and is therefore unnecessarily slow even when no memory allocation is necessary.

The solution we adopt is to tell the outside engine how big the vector has to be, and then each time the method is called, to return an unsigned long saying how many cash-flows have been generated. Thus we have two pure virtual methods:

```
virtual unsigned long MaxNumberOfCashFlows() const=0;
virtual unsigned long CashFlows(const MJArray& SpotValues,
                                std::vector<CashFlow>&
                                GeneratedFlows) const=0;
```

So in summary our objects will communicate as follows:

(i) The path generator asks the product what times it needs spot for, and it passes back an array.

(ii) The accounting part of the engine asks the product what cash-flow times are possible, and it passes back an array. The engine then computes all the possible discount factors.

(iii) The accounting part of the engine asks the product the maximum number of cash flows it can generate, and sets up a vector of that size.

(iv) For each path, the engine gets an array of spot values from the path generator.

(v) The array of spot values is passed into the product, which passes back the number of cash-flows, and puts their values into the vector.

(vi) The cash-flows are discounted appropriately and summed, and the total value is passed into the statistics gatherer.

(vii) After all the looping is done, the final results are obtained from the statistics gatherer.

7.4 The base classes

Having discussed in previous sections what classes will be needed and how they should communicate, in this section we give the implementations of the base classes.

In PathDependent.h, we define the CashFlow and the PathDependent classes which give our path-dependent exotic option.

Listing 7.1 (PathDependent.h)

```cpp
#ifndef PATH_DEPENDENT_H
#define PATH_DEPENDENT_H
#include <Arrays.h>
#include <vector>

class CashFlow
{
public:
    double Amount;
    unsigned long TimeIndex;

    CashFlow(unsigned long TimeIndex_=0UL, double Amount_=0.0)
                : TimeIndex(TimeIndex_),
                  Amount(Amount_){};
};

class PathDependent
{
public:
    PathDependent(const MJArray& LookAtTimes);

    const MJArray& GetLookAtTimes() const;

    virtual unsigned long MaxNumberOfCashFlows() const=0;
    virtual MJArray PossibleCashFlowTimes() const=0;
    virtual unsigned long CashFlows(const MJArray& SpotValues,
                                    std::vector<CashFlow>&
                                        GeneratedFlows) const=0;
    virtual PathDependent* clone() const=0;

    virtual ~PathDependent(){}
private:
    MJArray LookAtTimes;
};
#endif
```

The CashFlow class is really just a struct as it has public data members. Note that we ensure that it has a default constructor by giving the constructor default arguments, this is necessary in order to use it with STL containers which need a default constructor for certain operations such as creating a vector of arbitrary size.

The base class for PathDependent really does not do much except define the interface. We have made the base class store the LookAtTimes as every possible product will need these times, and provided the method GetLookAtTimes to obtain them. As usual, we include a clone method for virtual copy construction, and a virtual destructor to make sure that there are no memory leaks arising from destroying base class objects instead of inherited ones.

The source code is suitably short:

Listing 7.2 (PathDependent.cpp)

```
#include <PathDependent.h>

PathDependent::PathDependent(const MJArray& LookAtTimes_)
                        :    LookAtTimes(LookAtTimes_)
{}

const MJArray& PathDependent::GetLookAtTimes() const
{
    return LookAtTimes;
}
```

There is a bit more to the base class for the engine as it will actually handle the accounting for the cash-flows.

Listing 7.3 (ExoticEngine.h)

```
#ifndef EXOTIC_ENGINE_H
#define EXOTIC_ENGINE_H
#include <wrapper.h>
#include <Parameters.h>
#include <PathDependent.h>
#include <MCStatistics.h>
#include <vector>
class ExoticEngine
{
public:
    ExoticEngine(const Wrapper<PathDependent>&
                    The Product_, const Parameters& r_);

    virtual void GetOnePath(MJArray& SpotValues)=0;
```

```
    void DoSimulation(StatisticsMC& TheGatherer,
                            unsigned long NumberOfPaths);
    virtual ~ExoticEngine(){}
    double DoOnePath(const MJArray& SpotValues) const;

private:
    Wrapper<PathDependent> TheProduct;
    Parameters r;
    MJArray Discounts;
    mutable std::vector<CashFlow> TheseCashFlows;
};
#endif
```

The engine has four data members. The product is stored using the Wrapper template as we do not know its type. The interest rates are stored using the Parameters class which will allow us variable ones if we so desire. We also delegate computation of integrals to the Parameters class, and not have to worry about them here.

We have an array Discounts, which will be used to store the discount factors in order for the possible cash-flow times. Finally we have a mutable data member TheseCashFlows. This means that it can change value inside const member functions. The idea is that the variable is not really a data member, but instead it is a workspace variable: this it is faster to declare once and for all in the class definition. Remember that creating and destroying containers can be time-consuming so we design the class so that the vector is created once and for all at the beginning.

Note that we split our main method; it has two auxiliary methods, DoOnePath and GetOnePath. The second of these is pure virtual and therefore will be defined in an inherited class which will involve a choice of stochastic process and model. Note that this method is not constant as we will want a different set of spot values every time, and so it will necessarily change something about the state of the object. The other of the methods does everything necessary for one path given the spot values. This is const as turning spot values into prices is a purely functional action with no underlying changes. Both these methods pass arrays by reference in order to avoid any memory allocation. Note the implicit assumption that the array passed into GetOnePath is of the correct size.

The source code for implementing the base class is fairly simple and straightforward as all the hard work has been hived off into auxiliary classes.

Listing 7.4 (ExoticEngine.cpp)

```
#include <ExoticEngine.h>
#include <cmath>
```

```
ExoticEngine::ExoticEngine(const Wrapper<PathDependent>&
                                   TheProduct_,
                             const Parameters& r_)

                     :
                 TheProduct(TheProduct_),
                 r(r_),
                 Discounts(TheProduct_->PossibleCashFlowTimes())
{
    for (unsigned long i=0; i < Discounts.size(); i++)
        Discounts[i] = exp(-r.Integral(0.0, Discounts[i]));

    TheseCashFlows.resize(TheProduct->MaxNumberOfCashFlows());
}


void ExoticEngine::DoSimulation(StatisticsMC& TheGatherer,
                                  unsigned long NumberOfPaths)
{
    MJArray SpotValues(TheProduct->GetLookAtTimes().size());

    TheseCashFlows.resize(TheProduct->MaxNumberOfCashFlows());
    double thisValue;

    for (unsigned long i =0; i < NumberOfPaths; ++i)
    {
        GetOnePath(SpotValues);
        thisValue = DoOnePath(SpotValues);
        TheGatherer.DumpOneResult(thisValue);
    }

    return;
}


double ExoticEngine::DoOnePath(const MJArray&
                                     SpotValues) const
{
    unsigned long NumberFlows =
            TheProduct->CashFlows(SpotValues,
                                    TheseCashFlows);
    double Value=0.0;
```

```
for (unsigned i =0; i < NumberFlows; ++i)
    Value += TheseCashFlows[i].Amount *
        Discounts[TheseCashFlows[i].TimeIndex];

    return Value;
}
```

The constructor stores the inputs, computes the discount factors necessary, and makes sure the cash-flows vector is of the correct size. The DoSimulation method loops through all the paths, calling GetOnePath to get the array of spot value and then passes them into DoOnePath to get the value for that set of spot values. This value is then dumped into the statistics gatherer.

DoOnePath is only slightly more complicated. The array of spot values is passed into the product to get the cash-flows. These cash-flows are then looped over and discounted appropriately. The discounting is simplified by using the precomputed discount factors.

We have now set up the structure for pricing path-dependent exotic derivatives but we still have to actually define the classes which will do the path generation and define the products.

7.5 A Black–Scholes path generation engine

The Black–Scholes engine will produce paths from the risk-neutral Black–Scholes process. The paths will be an array of spot values at the times specified by the product. We allow the possibility of variable interest rates and dividend rates, as well as variable but deterministic volatility. The stock price therefore follows the process

$$dS_t = (r(t) - d(t))S_t dt + \sigma(t)S_t dW_t, \qquad (7.1)$$

with S_0 given. To simulate this process at times $t_0, t_1, \ldots, t_{n-1}$, we need n independent $N(0, 1)$ variates W_j and we set

$$\log S_{t_0} = \log S_0 + \int_0^{t_0} \left(r(s) - d(s) - \frac{1}{2}\sigma(s)^2 \right) ds + \sqrt{\int_0^{t_0} \sigma(s)^2 ds}\, W_0, \qquad (7.2)$$

and put

$$\log S_{t_j} = \log S_{t_{j-1}} + \int_{t_{j-1}}^{t_j} \left(r(s) - d(s) - \frac{1}{2}\sigma(s)^2 \right) ds + \sqrt{\int_{t_{j-1}}^{t_j} \sigma(s)^2 ds}\, W_j. \qquad (7.3)$$

We implement this procedure in `ExoticBSEngine.h` and `ExoticBS-Engine.cpp`.

Listing 7.5 (ExoticBSEngine.h)

```
#ifndef EXOTIC_BS_ENGINE_H
#define EXOTIC_BS_ENGINE_H
#include <ExoticEngine.h>
#include <Random2.h>

class ExoticBSEngine : public ExoticEngine
{
public:
    ExoticBSEngine(const Wrapper<PathDependent>& TheProduct_,
                   const Parameters& R_,
                   const Parameters& D_,
                   const Parameters& Vol_,
                   const Wrapper<RandomBase>& TheGenerator_,
                   double Spot_);

    virtual void GetOnePath(MJArray& SpotValues);
    virtual ~ExoticBSEngine(){}

private:
    Wrapper<RandomBase> TheGenerator;
    MJArray Drifts;
    MJArray StandardDeviations;
    double LogSpot;
    unsigned long NumberOfTimes;
    MJArray Variates;
};
#endif
```

Listing 7.6 (ExoticBSEngine.cpp)

```
#include <ExoticBSEngine.h>
#include <cmath>

void ExoticBSEngine::GetOnePath(MJArray& SpotValues)
{
    TheGenerator->GetGaussians(Variates);
```

```
    double CurrentLogSpot = LogSpot;

    for (unsigned long j=0; j < NumberOfTimes; j++)
    {
        CurrentLogSpot += Drifts[j];
        CurrentLogSpot += StandardDeviations[j]*Variates[j];
        SpotValues[j] = exp(CurrentLogSpot);
    }

    return;
}


ExoticBSEngine::ExoticBSEngine(const Wrapper<PathDependent>&
                                                 TheProduct_,
                        const Parameters& R_,
                        const Parameters& D_,
                        const Parameters& Vol_,
                        const Wrapper<RandomBase>&
                                    TheGenerator_,
                        double Spot_)
                        :
                        ExoticEngine(TheProduct_,R_),
                        TheGenerator(TheGenerator_)
{
    MJArray Times(TheProduct_->GetLookAtTimes());
    NumberOfTimes = Times.size();

    TheGenerator->ResetDimensionality(NumberOfTimes);
    Drifts.resize(NumberOfTimes);
    StandardDeviations.resize(NumberOfTimes);

    double Variance = Vol_.IntegralSquare(0,Times[0]);

    Drifts[0] = R_.Integral(0.0,Times[0])
            - D_.Integral(0.0,Times[0]) - 0.5 * Variance;
    StandardDeviations[0] = sqrt(Variance);

    for (unsigned long j=1; j < NumberOfTimes; ++j)
    {
```

```
    double thisVariance =
        Vol_.IntegralSquare(Times[j-1],Times[j]);
    Drifts[j] = R_.Integral(Times[j-1],Times[j])
                - D_.Integral(Times[j-1],Times[j])
                - 0.5 * thisVariance;
    StandardDeviations[j] = sqrt(thisVariance);
}


LogSpot = log(Spot_);
Variates.resize(NumberOfTimes);
}
```

The integrals and square-roots are the same for every path and so can be precomputed. The constructor therefore gets the times from the product, and uses them to compute the integrals of the drifts and the standard deviations which are stored as data members. Note that the class does not bother to store the times as it is only the constructor which needs to know what they are. In any case, the product is passed up to the base class and it could be retrieved from there if it were necessary.

The generation will of course require a random number generator and we pass in a wrapped RandomBase object to allow us to plug in any one we want without having to do any explicit memory handling. We have a data member Variates so that the array can be defined once and for all at the beginning: once again this is with the objective of avoiding unnecessary creation and deletion of objects. We store the log of the initial value of spot as this is the most convenient for carrying out the path generation.

As we have done a lot of precomputation in the constructor, the routine to actually generate a path is fairly simple. We simply get the variates from the generator and loop through the times. For each time, we add the integrated drift to the log, and then add the product of the random number and the standard deviation. To minimize the number of calls to log and exp, we keep track of the log of the spot at all times, and convert into spot values as necessary. We thus have NumberOfTimes calls to exp each path and no calls to log. As we will have to exponentiate to change our Gaussian into a log-normal variate at some point, this appears to be optimal for this design. If we were really worried that too much time was being spent on computing exponentials, one solution would be to change the design and pass the log of the values of spot back, and then pass these log values into the product. The product would then have the obligation to exponentiate them if necessary. For certain products such as a geometric Asian option this might well be faster as it would only involve one exponentiation instead of many. The main downside would be that for certain processes, such as a normal process or displaced diffusion, one might end up having to take unnecessary logs.

7.6 An arithmetic Asian option

Before we can run our engine, we need one last thing, namely a concrete product to put in it. One simple example is an arithmetic Asian option. Rather than define a different class for each sort of pay-off, we use the already developed PayOff class as a data member.

The header file for the class is quite simple:

Listing 7.7 (PathDependentAsian.h)

```
#ifndef PATH_DEPENDENT_ASIAN_H
#define PATH_DEPENDENT_ASIAN_H


#include <PathDependent.h>
#include <PayOffBridge.h>


class PathDependentAsian : public PathDependent
{
public:
    PathDependentAsian(const MJArray& LookAtTimes_,
                       double DeliveryTime_,
                       const PayOffBridge& ThePayOff_);

    virtual unsigned long MaxNumberOfCashFlows() const;
    virtual MJArray PossibleCashFlowTimes() const;
    virtual unsigned long CashFlows(const MJArray& SpotValues,
                std::vector<CashFlow>& GeneratedFlows) const;
    virtual ~PathDependentAsian(){}
    virtual PathDependent* clone() const;
private:
    double DeliveryTime;
    PayOffBridge ThePayOff;
    unsigned long NumberOfTimes;
};
#endif
```

The methods defined are just the ones required by the base class. We pass in the averaging times as an array and we provide a separate delivery time to allow for the possibility that the pay-off occurs at some time after the last averaging date. Note that the use of `PayOffBridge` class means that the memory handling is handled internally, and this class does not need to worry about assignment, copying and destruction.

The source file is fairly simple too.

Listing 7.8 (PathDependentAsian.cpp)

```
#include <PathDependentAsian.h>

PathDependentAsian::PathDependentAsian(const MJArray&
                                            LookAtTimes_,
                          double DeliveryTime_,
                          const PayOffBridge&ThePayOff_)
                          :
                          PathDependent(LookAtTimes_),
                          DeliveryTime(DeliveryTime_),
                          ThePayOff(ThePayOff_),
                          NumberOfTimes(LookAtTimes_.size())
{
}


unsigned long PathDependentAsian::MaxNumberOfCashFlows() const
{
    return 1UL;
}


MJArray PathDependentAsian::PossibleCashFlowTimes() const
{
    MJArray tmp(1UL);
    tmp[0] = DeliveryTime;
    return tmp;
}


unsigned long PathDependentAsian::CashFlows(const MJArray&
                                            SpotValues,
             std::vector<CashFlow>& GeneratedFlows) const
{
    double sum = SpotValues.sum();
    double mean = sum/NumberOfTimes;

    GeneratedFlows[0].TimeIndex = 0UL;
```

```
    GeneratedFlows[0].Amount = ThePayOff(mean);

    return 1UL;
}
```

```
PathDependent* PathDependentAsian::clone() const
{
    return new PathDependentAsian(*this);
}
```

Note that our option only ever returns one cash-flow so the maximum number of cash-flows is 1. It only ever generates cash-flows at the delivery time so the `PossibleCashFlowTimes` method is straightforward too. The `CashFlows` method takes the spot values, sums them, divides by the number of them and calls `ThePay-Off` to find out what the pay-off is. The answer is then written into the `Generated-Flows` array and we are done.

7.7 Putting it all together

We now have everything we need to price an Asian option. We give an example of a simple interface program in `EquityFXMain.cpp`.

Listing 7.9 (`EquityFXMain.cpp`)

```
/*
    uses source files
    AntiThetic.cpp
    Arrays.cpp,
    ConvergenceTable.cpp,
    ExoticBSEngine.cpp
    ExoticEngine.cpp
    MCStatistics.cpp
    Normals.cpp
    Parameters.cpp,
    ParkMiller.cpp,
    PathDependent.cpp
    PathDependentAsian.cpp
    PayOff3.cpp,
    PayOffBridge.cpp,
    Random2.cpp,
*/
```

```cpp
#include<ParkMiller.h>
#include<iostream>
using namespace std;
#include<MCStatistics.h>
#include<ConvergenceTable.h>
#include<AntiThetic.h>
#include<PathDependentAsian.h>
#include<ExoticBSEngine.h>
int main()
{
    double Expiry;
    double Strike;
    double Spot;
    double Vol;
    double r;
    double d;
    unsigned long NumberOfPaths;
    unsigned NumberOfDates;

    cout << "\nEnter expiry\n";
    cin >> Expiry;

    cout << "\nStrike\n";
    cin >> Strike;

    cout << "\nEnter spot\n";
    cin >> Spot;

    cout << "\nEnter vol\n";
    cin >> Vol;

    cout << "\nr\n";
    cin >> r;

    cout << "\nd\n";
    cin >> d;

    cout << "Number of dates\n";
    cin >> NumberOfDates;
```

```
cout << "\nNumber of paths\n";
cin >> NumberOfPaths;


PayOffCall thePayOff(Strike);


MJArray times(NumberOfDates);


for (unsigned long i=0; i < NumberOfDates; i++)
    times[i] = (i+1.0)*Expiry/NumberOfDates;


ParametersConstant VolParam(Vol);
ParametersConstant rParam(r);
ParametersConstant dParam(d);


PathDependentAsian theOption(times, Expiry, thePayOff);


StatisticsMean gatherer;
ConvergenceTable gathererTwo(gatherer);


RandomParkMiller generator(NumberOfDates);


AntiThetic GenTwo(generator);


ExoticBSEngine theEngine(theOption, rParam, dParam,
                         VolParam, GenTwo, Spot);


theEngine.DoSimulation(gathererTwo, NumberOfPaths);


vector<vector<double> > results =
    gathererTwo.GetResultsSoFar();


cout <<"\nFor the Asian call price the results are \n";

{
for (unsigned long i=0; i < results.size(); i++)
    {
    for (unsigned long j=0; j < results[i].size(); j++)
        cout << results[i][j] << " ";

    cout << "\n";
```

```
    }}

    double tmp;
    cin >> tmp;

    return 0;
}
```

7.8 Key points

In this chapter, we saw how we can put the ideas developed in the previous chapters together to build a pricer for exotic options.

- An important part of the design process is identifying the necessary components and specifying how they talk to each other.
- The template pattern involves deferring the implementation of an important part of an algorithm to an inherited class.
- If an option class knows nothing that is not specified in the term-sheet then it is much easier to reuse.
- We can reuse the PayOff class to simplify the coding of our more complicated path-dependent derivatives.

7.9 Exercises

Exercise 7.1 Write a class to do geometric Asian options.

Exercise 7.2 Write a class to do discrete knock-out options that pay a rebate at the time of rebate.

Exercise 7.3 Rewrite the classes here so that they pass the logs of spot values around instead of the spot values. Show that the discrete barrier option and the geometric Asian need fewer exponentiations.

Exercise 7.4 Implement an engine for pricing when the spot price is normal instead of log-normal.

Exercise 7.5 Write a class that pays the difference in pay-offs of two arbitrary path-dependent derivatives.

8

Trees

8.1 Introduction

We have studied Monte Carlo code in detail: now we examine how we can apply similar techniques to pricing on trees. Before we can start designing the code, we need to fix the underlying mathematics. The point of view we adopt is that a tree is a method of approximating the risk-neutral expectation. In particular, we assume that we are pricing a derivative on a stock following geometric Brownian motion with a constant volatility σ. We let the continuously compounding interest rate be r and the continuous dividend rate be d. The dynamics for the stock in the risk-neutral measure are therefore

$$dS = (r - d)Sdt + \sigma S dW_t.$$ (8.1)

The value of a European option with expiry T is then

$$e^{-rT}\mathbb{E}(C(S, T)),$$ (8.2)

where $C(S, T)$ is the final pay-off.

When we price on a binomial tree, we divide time into steps and across each step we assume that the underlying Brownian motion can only move a fixed amount up or down. The dynamics of the stock price under geometric Brownian motion are such that

$$S_t = S_0 e^{(r-d-\frac{1}{2}\sigma^2)t+\sigma W_t}.$$ (8.3)

We wish to discretize W_t. We have N steps to get from 0 to T. Each time step is therefore of length T/N. Across step l, we need to approximate

$$W_{(l+1)T/N} - W_{lT/N} = \sqrt{\frac{T}{N}}N(0, 1).$$ (8.4)

There is only one random variable taking precisely two values which has the same mean and variance as $N(0, 1)$, and this variable takes ± 1 with probability $1/2$. We therefore take a set of N independent random variables X_j with this distribution, and approximate $W_{lT/N}$ by

$$Y_l = \sqrt{\frac{T}{N}} \sum_{j=1}^{l} X_j. \tag{8.5}$$

The approximation for $S_{lT/N}$ is then

$$S_0 e^{(r-d-\frac{1}{2}\sigma^2)lT/N + \sigma Y_l}.$$

Note the crucial point here that since the value of S_t does not depend on the path of W_t but solely upon its value at time t, it is only the value of Y_l that matters not the value of each individual X_j. This is crucial because it means that our tree is recombining; it does not matter whether we go down then up, or up then down. This is not the case if we allow variable volatility, which is why we have assumed its constancy.

The nature of martingale pricing means that the value at a given time-step is equal to the discounted value at the next time-step, thus if we let $S_{l,k}$ be the value of the stock at time Tl/N if Y_l is k, we have that

$$C(S_{l,k}, Tl/N) = e^{-rT/N} \mathbb{E}(S_{l+1}(Y_{l+1}|Y_l = k)),$$
$$= \frac{1}{2} e^{-rT/N} \Big(C\big(S_{l,k} e^{(r-d-\frac{1}{2}\sigma^2)T/N + \sigma\sqrt{T/N}}, (l+1)T/N\big)$$
$$+ C\big(S_{l,k} e^{(r-d-\frac{1}{2}\sigma^2)T/N - \sigma\sqrt{T/N}}, (l+1)T/N\big)\Big). \tag{8.6}$$

Note that we are not doing true martingale pricing in the discrete world in that we are not adjusting probabilities to make sure the assets are discrete martingales; we are instead approximating the continuous martingales with discrete random variables which are almost, but not quite, martingales.

What does all this buy us? The value of $C(S_{N,k})$ is easy to compute for any value of k: we just plug the stock price into the final pay-off. Since we have a formula that expresses the value at step l into that at step $l+1$, we can now just backwards iterate through the tree in order to get the price at time zero.

However, the purpose of a tree is not to price a European option; there are lots of better ways of doing that, including analytical solutions and numerical integration. The reason trees were introduced was that they give an effective method for pricing American options. The analysis for an American option is similar except that the value at a point in the tree is the maximum of the exercise value at that point and the discounted expected value at the next time. This corresponds to the optimal strategy of exercise if and only if exercise gives more money than not exercising.

Our algorithm for pricing an American option is therefore as follows:

(i) Create an array of final spot values which are of the form

$$S_0 e^{(r-d-\frac{1}{2}\sigma^2)T+\sigma\sqrt{T/N}j}$$

where j ranges from $-N$ to N.

(ii) For each of these spot values evaluate the pay-off and store it.

(iii) At the previous time-slice compute the possible values of spot: these will be of the form

$$S_0 e^{(r-d-\frac{1}{2}\sigma^2)(N-1)T/N+\sigma\sqrt{T/N}j},$$

where j ranges from $-(N-1)$ to $N-1$.

(iv) For each of these values of spot, compute the pay-off and take the maximum with the discounted pay-off of the two possible values of spot at the next time.

(v) Repeat 3,4 until time zero is reached.

What else could we price on a tree? We could do a barrier option or an American option that could only be exercised within certain time periods. For a knock-out barrier option, the procedure would be the same as for the European, except that the value at a point in the tree would be zero if it lay behind the barrier. For an American option with limited early exercise the procedure would be the same again, except that we would only take the maximum at times at which early exercise was allowed. So in each case, when we change the option, all that alters is the rule for updating the value at a point in the tree.

Note that in our formulation, we have not used any no-arbitrage arguments. The reason is that we have implicitly assumed that the no-arbitrage arguments have been done in the continuous case before any discretization has been carried out. This means that the completeness property of a binomial tree, i.e. that it generates a *unique* no-arbitrage price, is not relevant. In particular, we could replace X_j by any random variable with mean 0 and variance 1. If we use X'_j which takes values $-\sqrt{2}, 0, \sqrt{2}$ with probabilities $0.25, 0.5$ and 0.25 respectively, then we obtain a trinomial tree on which we could carry out a very similar analysis and algorithm. We could in fact go much further and have as many values as we wanted as long as we took care to make sure that the tree still recombined.

8.2 The design

Having re-examined the mathematics and the algorithms, we are now in a position to think about the design. Here are some concepts that our discussion has thrown up:

- the discretization;
- the final pay-off of the option;

- the rule for deciding the value of an option at a point in the tree given spot and the discounted future value of the option.

The first of these concepts decides the shape of the tree, whereas the second and third are properties of the option. There is thus an obvious orthogonalization: we have a tree class which handles the discretization, and a second class which deals with the final pay-off and the rule at previous times. In fact, we have already developed a class, `PayOff`, to encapsulate vanilla option pay-offs and it is ideal for reuse here.

There are a number of ways we could approach the second class. We could inherit from `PayOffBridged` since we could view our class as adding structure to an existing class. Whilst this would work in code, I personally dislike it as an option being priced on a tree is not a type of pay-off, and so the inheritance is not expressing an *is a* relationship. Another approach might be simply to define a second class to do the calculation rule, and plug both the final pay-off and the calculation rule into the tree. Since for American options the final pay-off is generally relevant at all times, such an approach seems sub-optimal as it might require two copies of the pay-off.

Ultimately, the pay-off is an aspect of the option, and it therefore makes more sense to define it as data member of the class which can referenced via a final pay-off method. Thus we define options on trees via an abstract base class which has three defining methods:

- `FinalTime` indicates the expiry time of the option;
- `FinalPayOff` gives the final pay-off as a function of spot;
- `PreFinalValue` gives the value at a point in the tree as a function of spot, time and the discounted future value of the option.

Note that by defining the option class in this fashion, we have not allowed it to know anything about interest rates nor the process of the underlying. This means it can be used in any tree-like structure provided the structure is always in a position to let it know its own discounted future value. Note the difference here between the option classes we are defining here and those we defined for Monte Carlo: whilst we are trying to encapsulate similar concepts, the difference is in the information we are able to feed in at a given time. For Monte Carlo, the entire history of spot is easy but the discounted future value of an option is hard, whereas on a tree the discounted future value is easy but the history is hard. However, both classes have easy access to the pay-off which means we are able to share the pay-off class.

Our other concept is the tree itself. The tree really has two aspects: the placement of the nodes of the tree and the computing of the option value at each of the nodes. Whilst we could further orthogonalize and define separate classes for each of these, we write a single class to do the binomial tree which takes in an option as an argument. An important point to note is that as the placement of the nodes does not depend upon the option, we can save ourselves time if we want to price several options by placing the nodes once and then pricing all the options on the same tree.

Given this fact, we design our tree class in such a way that the tree is built once, and then any option can be valued on the tree via a separate method.

8.3 The `TreeProduct` class

As we have decided to model the tree and the product separately, we develop a class hierarchy for the products we can value on trees. As usual, we use an abstract base class to define an interface. We define the class, `TreeProduct`, in `TreeProducts.h`.

Listing 8.1 (`TreeProducts.h`)

```
#ifndef TREE_PRODUCTS_H
#define TREE_PRODUCTS_H

class TreeProduct
{
public:
    TreeProduct(double FinalTime_);
    virtual double FinalPayOff(double Spot) const=0;
    virtual double PreFinalValue(double Spot,
                                 double Time,
                                 double DiscountedFutureValue)
                                const=0;
    virtual ~TreeProduct(){}
    virtual TreeProduct* clone() const=0;
    double GetFinalTime() const;

private:
    double FinalTime;
};
#endif
```

The only data member for the base class is `FinalTime`, and we provide a `GetFinalTime` to allow its value to be read. Note that this places the constraint on our products that they actually have a time of expiry! Thus we are implicitly disallowing perpetual options, however this is a good thing as it is not clear how to go about valuing such an option using a tree. Ultimately, one would have to approximate using a product with a finite expiry and it is probably better to do so explicitly than implicitly.

We provide the usual `clone` method and a virtual destructor to allow virtual copying, and to ensure the absence of memory leaks after virtual copying. The remaining methods are pure virtual and specify the value of the product at expiry and at previous times as we discussed above.

As most of the base class is abstract the source file is short:

Listing 8.2 (`TreeProducts.cpp`)

```
#include <TreeProducts.h>


TreeProduct::TreeProduct(double FinalTime_)
: FinalTime(FinalTime_)
{
}

double TreeProduct::GetFinalTime() const
{
    return FinalTime;
}
```

We provide below two concrete implementations of tree products: the European option and the American option. As we specify the pay-off using the `PayOff-Bridged` class, we do not need to write separate classes for calls and puts. The header files are straightforward, the only changes from the base class being the addition of a data member to specify the pay-off and the fact that virtual methods are now concrete instead of abstract. Note that everything to do with the expiry time is taken care of by the base class; the constructors need only pass it on to the base class.

Listing 8.3 (`TreeAmerican.h`)

```
#ifndef TREE_AMERICAN_H
#define TREE_AMERICAN_H

#include <TreeProducts.h>
#include <PayOffBridge.h>

class TreeAmerican : public TreeProduct
{

public:
    TreeAmerican(double FinalTime,
                 const PayOffBridge& ThePayOff_);
```

```
    virtual TreeProduct* clone() const;
    virtual double FinalPayOff(double Spot) const;
    virtual double PreFinalValue(double Spot,
                                 double Time,
                                 double DiscountedFutureValue)
                                const;
    virtual ~TreeAmerican(){}

private:
    PayOffBridge ThePayOff;

};
#endif
```

The header for the European option is very similar:

Listing 8.4 (TreeEuropean.h)

```
#ifndef TREE_EUROPEAN_H
#define TREE_EUROPEAN_H

#include <TreeProducts.h>
#include <PayOffBridge.h>

class TreeEuropean : public TreeProduct
{

public:
    TreeEuropean(double FinalTime,
                 const PayOffBridge& ThePayOff_);

    virtual TreeProduct* clone() const;
    virtual double FinalPayOff(double Spot) const;
    virtual double PreFinalValue(double Spot,
                                 double Time,
                                 double DiscountedFutureValue)
                                const;
    virtual ~TreeEuropean(){}

private:
    PayOffBridge ThePayOff;
};
#endif
```

The source files are also straightforward.

Listing 8.5 (TreeAmerican.cpp)

```cpp
#include <TreeAmerican.h>
#include <minmax.h>


TreeAmerican::TreeAmerican(double FinalTime,
                            const PayOffBridge& ThePayOff_)
                : TreeProduct(FinalTime),
                  ThePayOff(ThePayOff_)
{
}


TreeProduct* TreeAmerican::clone() const
{
    return new TreeAmerican(*this);
}


double TreeAmerican::FinalPayOff(double Spot) const
{
    return ThePayOff(Spot);
}


double TreeAmerican::PreFinalValue(double Spot,
                        double ,
// Borland compiler doesnt like unused named variables
                    double DiscountedFutureValue) const
{
    return max(ThePayOff(Spot), DiscountedFutureValue);
}
```

and

Listing 8.6 (TreeEuropean.cpp)

```cpp
#include <TreeEuropean.h>
#include <minmax.h>


TreeEuropean::TreeEuropean(double FinalTime,
                            const PayOffBridge& ThePayOff_)
                : TreeProduct(FinalTime),
                  ThePayOff(ThePayOff_)
```

```
{
}


double TreeEuropean::FinalPayOff(double Spot) const
{
    return ThePayOff(Spot);
}


double TreeEuropean::PreFinalValue(double,
               //Spot,  Borland compiler
                          double,
            //Time,   doesnt like unused named variables
             double DiscountedFutureValue) const
{
    return DiscountedFutureValue;
}


TreeProduct* TreeEuropean::clone() const
{
    return new TreeEuropean(*this);
}
```

The implementations of the methods are very simple – all the pay-offs are sub-contracted to the PayOffBridged class in any case. For the European option at an interior node, the rule for computing the PreFinalValue is very simple: just return the discounted future value that was input. For the American option, it is only slightly harder; we take the maximum with the intrinsic value.

Note the slight oddity that we do not name the unused variables Spot and Time in the PreFinalValue method. This is because some compilers issue a warning if a variable is named but not used, to ensure that variables are not left unused accidentally.

8.4 A tree class

We give a simple implementation of a binomial tree class in this section. We design the tree to work in three pieces. The constructor does little except store the parameters. The BuildTree method actually makes the tree. In particular, it computes the locations of all the nodes, and the discounts needed to compute the expectations backwards in the tree. As it is not intended that the BuildTree method should be called from outside the class, we make it protected which allows the possibility that an inherited class may wish to use it without allowing any other external access.

The method which does the actual pricing is GetThePrice. Note that it takes in a TreeProduct by reference. As the argument is an abstract base class this means that an object from an inherited class must be passed in. Note that since we not need to store the object passed in, we do not need virtual constructors, wrappers or bridges. This method builds the tree if necessary, checks that the product has the right expiry time and then prices it. Our design is such that we can price multiple products with the same expiry; we call the method multiple times and only build the tree once. As we wish to be able to do this, we store the entire tree. Note that this is not really necessary since for any given time slice, we only need the next time slice and so we could easily save a lot of memory by only ever having two arrays defined. However, unless one is doing an awful lot of steps, memory will not be an issue, and this approach has the added benefit that if one wishes to analyze certain aspects of the product, such as where the exercise boundary lies for an American option, it is better to have the entire tree. We present the class in BinomialTree.h.

Listing 8.7 (BinomialTree.h)

```
#pragma warning( disable : 4786 )

#include <TreeProducts.h>
#include <vector>
#include <Parameters.h>
#include <Arrays.h>

class SimpleBinomialTree
{
public:
    SimpleBinomialTree(double Spot_,
                       const Parameters& r_,
                       const Parameters& d_,
                       double Volatility_,
                       unsigned long Steps,
                       double Time);

    double GetThePrice(const TreeProduct& TheProduct);

protected:
    void BuildTree();

private:
```

```
    double Spot;
    Parameters r;
    Parameters d;
    double Volatility;
    unsigned long Steps;
    double Time;
    bool TreeBuilt;

    std::vector<std::vector<std::pair<double, double> > >
        TheTree;
    MJArray Discounts;
};
```

Note that we store the tree as a vector of vectors of pairs of doubles. This is why
we have the #pragma at the start; without the pragma, we get a warning message
telling us that the debug info is too long (under Visual C++ 6.0).

A pair is a simple template class in the STL which simply gives a class with
two data members of the appropriate types. They are accessed as public members
as first and second. Note that an alternative implementation would be to have
two trees: one for the spot and another for option values. However, that would
require twice as much work when resizing, and more importantly one would then
have to be careful to ensure that spot and the associated option value were always
attached to the same indices. With the use of a pair, we get this for free.

We have allowed general Parameters for r and d: this is because variable interest and dividend rates change little in the analysis or construction of the tree. If
we can add extra functionality with little cost, why not do so? We do not, however,
allow variable volatility as it greatly complicates node placement; we would lose
the property that the spot price is a simple function of the underlying Brownian
motion.

We use a bool to indicate whether the tree has been built yet. We store the
number of steps as an unsigned long and this is fixed in the constructor. One
could easily add an additional method to allow the number of steps to be changed;
however one would gain little over just instantiating a new object with a different
number of steps. We have a data member for the discount factors needed so that
they need only be computed once in BuildTree.

We present the source code in BinomialTree.cpp.

Listing 8.8 (BinomialTree.cpp)

```
#include <BinomialTree.h>
#include <Arrays.h>
#include <cmath>
```

```
// the basic math functions should be in namespace std
// but aren't in VCPP6
#if !defined(_MSC_VER)
using namespace std;
#endif

SimpleBinomialTree::SimpleBinomialTree(double Spot_,
                                       const Parameters& r_,
                                       const Parameters& d_,
                                       double Volatility_,
                                       unsigned long Steps_,
                                       double Time_)
                                     : Spot(Spot_),
                                       r(r_),
                                       d(d_),
                                       Volatility(Volatility_),
                                       Steps(Steps_),
                                       Time(Time_),
                                       Discounts(Steps)
{
    TreeBuilt=false;
}

void SimpleBinomialTree::BuildTree()
{
  TreeBuilt = true;
  TheTree.resize(Steps+1);

    double InitialLogSpot = log(Spot);

  for (unsigned long i=0; i <=Steps; i++)
  {

      TheTree[i].resize(i+1);

      double thisTime = (i*Time)/Steps;

      double movedLogSpot =
            InitialLogSpot + r.Integral(0.0, thisTime)
                           - d.Integral(0.0, thisTime);
```

```
    movedLogSpot -=
            0.5*Volatility*Volatility*thisTime;

    double sd = Volatility*sqrt(Time/Steps);

    for (long j = -static_cast<long>(i), k=0;
                j <= static_cast<long>(i); j=j+2,k++)
            TheTree[i][k].first = exp(movedLogSpot+ j*sd);
  }

  for (unsigned long l=0; l <Steps; l++)
  {
    Discounts[l] =
      exp(- r.Integral(l*Time/Steps,(l+1)*Time/Steps));
  }
}

double SimpleBinomialTree::GetThePrice(const TreeProduct&
                                                    TheProduct)
{
  if (!TreeBuilt)
      BuildTree();

  if (TheProduct.GetFinalTime() != Time)
      throw("mismatched product in SimpleBinomialTree");

  for (long j = -static_cast<long>(Steps), k=0;
          j <=static_cast<long>( Steps); j=j+2,k++)
        TheTree[Steps][k].second =
            TheProduct.FinalPayOff(TheTree[Steps][k].first);

  for (unsigned long i=1; i <= Steps; i++)
  {
    unsigned long index = Steps-i;
    double ThisTime = index*Time/Steps;

    for (long j = -static_cast<long>(index), k=0;
    j <= static_cast<long>(index); j=j+2,k++)
    {
      double Spot = TheTree[index][k].first;
```

```
      double futureDiscountedValue = 0.5*Discounts[index]*
      (TheTree[index+1][k].second +
         TheTree[index+1][k+1].second);
      TheTree[index][k].second =
      TheProduct.PreFinalValue(Spot,ThisTime,
                                    futureDiscountedValue);
   }


  }
  return TheTree[0][0].second;
}
```

The code is fairly straightforward. The constructor initializes the basic class variables and does not do much else. The method BuildTree creates the tree. We resize the vector describing all the layers first. We then resize each layer so that it is of the correct size. Note that as the number of nodes in a layer grows with the number of steps, these inner vectors are all of different sizes. We then compute a basepoint for the layer in log-space which is just the zero Brownian motion point. We then loop through the nodes in each layer writing in the appropriate value of spot.

Note that as we are dealing with points in the tree corresponding to down moves we count using a long. This long has to be compared to the unsigned long i for the termination condition. We therefore have to be careful; if we simply compare these numbers a likely effect is that the routine will conclude that -1 is bigger than 1. Why? The compiler will implicitly convert the long into a large unsigned long. Clearly this is not the desired effect so we convert the unsigned long into a long before the comparison.

After setting up the tree, we also set up the array of Discounts which are product-independent, and therefore only need to be done once.

The main routine for actually doing the pricing, GetThePrice, is also straightforward. We have put in a test to make sure that the tree and product are compatible. This throws an error which is a simple string if they are not. Note that this means the program will terminate unless we have written a catch which is capable of catching the string.

We simply iterate backwards through the tree. First we compute the final layer using the FinalPayOff and write the values into the second element of each pair in the final layer. After this we simply iterate backwards, computing as we go. The final value of the product is then simply the value of the option at the single node in the first layer of the tree, and that is what we return.

8.5 Pricing on the tree

Having written all the classes we need to actually put them together to price something. We give a simple interface in TreeMain.cpp.

Listing 8.9 (TreeMain.cpp)

```
/*
requires
    Arrays.cpp
    BinomialTree.cpp
    BlackScholesFormulas.cpp
    Normals.cpp
    Parameters.cpp
    PayOff3.cpp
    PayOffBridge.cpp
    PayOffForward.cpp
    TreeAmerican.cpp
    TreeEuropean.cpp
    TreeProducts.cpp
*/
#include <BinomialTree.h>
#include <TreeAmerican.h>
#include <TreeEuropean.h>
#include <BlackScholesFormulas.h>
#include <PayOffForward.h>
#include <iostream>
using namespace std;
#include <cmath>
int main()
{

    double Expiry;
    double Strike;
    double Spot;
    double Vol;
    double r;
    double d;
    unsigned long Steps;
```

```
cout << "\nEnter expiry\n";
cin >> Expiry;

cout << "\nStrike\n";
cin >> Strike;

cout << "\nEnter spot\n";
cin >> Spot;

cout << "\nEnter vol\n";
cin >> Vol;

cout << "\nr\n";
cin >> r;

cout << "\nd\n";
cin >> d;

cout << "\nNumber of steps\n";
cin >> Steps;

PayOffCall thePayOff(Strike);

ParametersConstant rParam(r);
ParametersConstant dParam(d);

TreeEuropean europeanOption(Expiry,thePayOff);
TreeAmerican americanOption(Expiry,thePayOff);

SimpleBinomialTree theTree(Spot,rParam,dParam,Vol,Steps,
                           Expiry);
double euroPrice = theTree.GetThePrice(europeanOption);
double americanPrice = theTree.GetThePrice(americanOption);
cout << "euro price" << euroPrice << "amer price"
                     << americanPrice << "\n";

double BSPrice = BlackScholesCall(Spot,Strike,r,d,
                                  Vol,Expiry);
cout << "BS formula euro price" << BSPrice << "\n";
```

```
PayOffForward forwardPayOff(Strike);
TreeEuropean forward(Expiry,forwardPayOff);

double forwardPrice = theTree.GetThePrice(forward);
cout << "forward price by tree" << forwardPrice << "\n";

double actualForwardPrice =
    exp(-r*Expiry)*(Spot*exp((r-d)*Expiry)-Strike);
cout << "forward price" << actualForwardPrice << "\n";

Steps++; // now redo the trees with one more step
SimpleBinomialTree theNewTree(Spot,rParam,dParam,Vol,
                             Steps,Expiry);

double euroNewPrice =
    theNewTree.GetThePrice(europeanOption);
double americanNewPrice =
    theNewTree.GetThePrice(americanOption);

cout << "euro new price" << euroNewPrice
     << "amer new price" << americanNewPrice << "\n";

double forwardNewPrice = theNewTree.GetThePrice(forward);

cout << "forward price by new tree" << forwardNewPrice
                                    << "\n";

double averageEuro = 0.5*(euroPrice+euroNewPrice);
double averageAmer = 0.5*(americanPrice+americanNewPrice);
double averageForward = 0.5*(forwardPrice+forwardNewPrice);

cout << "euro av price" << averageEuro << "amer av price"
                        << averageAmer << "\n";
cout << "av forward" << averageForward << "\n";

double tmp;
cin >> tmp;

return 0;
}
```

To illustrate certain aspects of tree pricing, we price the European call, the American call and the forward. We then reprice them using one extra step. The reason we do this is that often pricing on trees gives rise to a zig-zag behaviour as nodes go above and below the money. The average of the price for two successive steps is therefore often a lot more accurate than a single price. We give the comparison Black–Scholes price for the European option as an assessment of accuracy. The code for the Black–Scholes price is in BlackScholesFormulas.h and BlackScholesFormulas.cpp; we discuss it in Appendix A.

A standard way of improving the accuracy of American option pricing is to use the European price as a control. As we know the true price of the European and we know the tree price also, we can assume that the American option has the same amount of error as the European and adjust its price accordingly. The principle here is the same as that for a control variate in Monte Carlo simulation.

We also give the price of a forward. Note that if we had required the discounted discretized process to be a martingale, then the forward would be priced absolutely correctly; however as it is only an approximation to a martingale rather than actually being one, the forward need not be precise.

As we have not defined a class for the forward's pay-off before, we define one in PayOffForward.h and PayOffForward.cpp.

Listing 8.10 (PayOffForward.h)

```
#ifndef PAY_OFF_FORWARD_H
#define PAY_OFF_FORWARD_H

#include <PayOff3.h>
class PayOffForward : public PayOff
{
public:
    PayOffForward(double Strike_);

    virtual double operator()(double Spot) const;
    virtual ~PayOffForward(){}
    virtual PayOff* clone() const;

private:
    double Strike;
};
#endif
```

Listing 8.11 (PayOffForward.cpp)

```
#include <PayOffForward.h>

double PayOffForward::operator () (double Spot) const
{
    return Spot-Strike;
}


PayOffForward::PayOffForward(double Strike_) : Strike(Strike_)
{
}


PayOff* PayOffForward::clone() const
{
    return new PayOffForward(*this);
}
```

The class is straightforward and the only difference from the class defined for the call is that we take spot minus strike, instead of the call pay-off.

8.6 Key points

In this chapter we have used the patterns developed earlier in the book to develop routines for pricing on trees.

- Tree pricing is based on the discretization of a Brownian motion.
- Trees are a natural way to price American options.
- On a tree, knowledge of discounted future values is natural but knowing about the past is not.
- We can re-use the pay-off class when defining products on trees.
- By having a separate class encapsulating the definition of a derivative on a tree, we can re-use the products for more general structures.
- European options can be used as controls for American options.

8.7 Exercises

We have developed a very simple tree and treated a couple of simple products. The approach here can easily be extended to many more cases. We suggest a few possibilities for the reader to try.

Exercise 8.1 Find a class that does barrier options in the same `TreeProduct` class hierarchy. Try it out. How stable is the price? How might you improve the stability?

Exercise 8.2 Develop a binomial tree for which the memory requirements grow linearly with the number of steps. How do the memory requirements grow for the class here?

Exercise 8.3 Write a trinomial tree class.

Exercise 8.4 Modify the code so that it will work under variable volatility. The key is to ensure that the integral of the square of the vol across each time step is the same. This means that the time steps will be of unequal length.

Exercise 8.5 Modify the tree so the implied stock price process makes the discounted price a martingale. Compare convergence for calls, puts and forwards.

Exercise 8.6 Implement an American knock-in option pricer on a tree. (Use an additional auxiliary variable to indicate whether or not the option has knocked-in, and compute the value at each node in both cases.)

9

Solvers, templates, and implied volatilities

9.1 The problem

Whatever model one is using to compute vanilla options' prices, it is traditional to quote prices in terms of the Black–Scholes implied volatility. The implied volatility is by definition the number to plug into the Black–Scholes formula to get the price desired. Thus we have the problem that we must solve the problem of finding the value σ such that

$$BS(S, K, r, d, T, \sigma) = \text{quoted price.}$$

In other words, we must invert the map

$$\sigma \mapsto BS(S, K, r, d, T, \sigma)$$

with the other parameters fixed.

The Black–Scholes formula is sufficiently complicated that there is no analytic inverse and this inversion must be carried out numerically. There are many algorithms for implementing such inversions; we shall study two of the simplest: bisection and Newton–Raphson. Our objective, as usual, is to illustrate the programming techniques for defining the interfaces in a reusable fashion rather than to implement the most efficient algorithms available. Indeed, we hope that the reader will combine the techniques here with algorithms found elsewhere, in for example [28], to produce robust and efficient reusable code.

Before proceeding to the coding and design issues, we recall the details of the aforementioned algorithms. Given a function, f, of one variable we wish to solve the equation

$$f(x) = y. \tag{9.1}$$

In the above, f is the Black–Scholes formula, x is volatility and y is the price. If

141

the function f is continuous, and for some a and b we have

$$f(a) < y, \tag{9.2}$$
$$f(b) > y, \tag{9.3}$$

then there must exist some c in the interval (a, b) such that $f(c) = x$. Bisection is one technique to find c. The idea is straightforward: we simply take the midpoint, m, of the interval, then one of three things must occur:

- $f(m) = y$ and we are done;
- $f(m) < y$ in which case there must be a solution in (m, b);
- $f(m) > y$ in which case there must be a solution in (a, m).

Thus by taking the midpoint, we either find the solution, or halve the size of the interval in which the solution exists. Repeating we must narrow in on the solution. In practice, we would terminate when we achieve

$$|f(m) - y| < \epsilon, \tag{9.4}$$

for some pre-decided tolerance, ϵ.

Bisection is robust but is not particularly fast. When we have a well-behaved function with an analytic derivative then Newton–Raphson can be much faster. The idea of Newton–Raphson is that we pretend the function is linear and look for the solution where the linear function predicts it to be. Thus we take a starting point, x_0, and approximate f by

$$g_0(x) = f(x_0) + (x - x_0) f'(x_0). \tag{9.5}$$

We have that $g_0(x)$ is equal to zero if and only if

$$x = \frac{y - f(x_0)}{f'(x_0)} + x_0. \tag{9.6}$$

We therefore take this value as our new guess x_1. We now repeat until we find that $f(x_n)$ is within ϵ of y.

Newton–Raphson is much faster than bisection provided we have an easily evaluated derivative. This is certainly the case for the Black–Scholes function. Indeed, for a call option the vega is easier to compute than the price is. However, as it involves passing two functions rather than one to a solver routine, it requires more sophisticated programming techniques to implement re-usably.

9.2 Function objects

We want to implement the bisection algorithm in a re-usable way; this means that we will need a way to pass the function f into the bisection routine. Since f may well be defined, as it is in our case, in terms of the value of a more complicated function with many parameters fixed, we will also need somehow to pass in the values of those auxiliary parameters.

There are, in fact, many different ways to tackle this problem. One method we have already studied is the engine template. With this approach, we define a base class for which the main method is to carry out the bisection. The main method calls a pure virtual method to get the value of $f(x)$. For any specific problem, we then define an inherited class which implements f appropriately. Whilst this method can work effectively, there are a couple of disadvantages. The first is that the function call is virtual which can lead to efficiency problems. There are two causes: the first is that to call a virtual function, the processor has to look up a virtual function table each time the function is called, and then jump to a location specified by the table. Clearly, it would be slightly faster not to have to look up the table. A more subtle and serious speed issue is that it is not possible to `inline` virtual functions. If the function is known beforehand, the compiler can inline it and eliminate the mechanics of the function call altogether. In addition, the compiler may be able to make additional optimizations as it sees all the code together at once. Whilst these speed issues are not particularly important whilst designing a solver, they are more critical when writing a class for other numerical routines such as numeric integration where often a large of calls are made to a function which is fast to evaluate. We therefore wish to develop a pattern which can be used in those contexts too.

The second disadvantage of inheriting from a solver base class is that it inhibits other inheritance. If we wish to inherit the class defining our function from some other class, we cannot inherit from the solver class as well without using multiple inheritance. Of course, one could use multiple inheritance but it tends to be tricky to get it to work in a bug-free fashion and I therefore tend to avoid it.

Having decided that we want to be able to input a function to our routine without using virtual functions, what other options do we have? One solution would be to use a function pointer but this would buy us little (if anything) over virtual functions. Another approach is templatization. The crucial point is that with templatization the type of the function being used in the optimization is decided at compile time rather than at runtime. This means that the compiler can carry out optimizations and inlining that depend on the type of the function since that information is now available to it.

The approach we adopt for specifying the function we wish to optimize uses the *function object*. We first encountered function objects in Section 2.1 when defining pay-offs. Recall that a function object is by definition an object for which `operator()` is defined. So if we have an object f of a class T for which

```
const operator()( double x) const
```

has been defined it is then legitimate to write f(y) for a double y, and this is equivalent to

f.operator()(y).

Thus our object f can be used with function-like syntax. However, as f is an object it can contain extra information. Thus if we want to solve for the implied volatility, the function object will take the volatility as an argument, but will also have, as extra parameters already stored, the values of r, d, T, S and K.

We thus obtain the class defined in BSCallClass.h:

Listing 9.1

```
//
//   BSCallClass.h
//

#ifndef BS_CALL_CLASS_H
#define BS_CALL_CLASS_H
class BSCall
{

public:

        BSCall(double r_, double d_,
                double T, double Spot_,
                double Strike_);

        double operator()(double Vol) const;


private:

        double r;
        double d;
        double T;
        double Spot;
        double Strike;

};
#endif
```

The source file is simple:

Listing 9.2

```
//
//   BSCallClass.cpp
//

#include <BSCallClass.h>
#include <BlackScholesFormulas.h>

BSCall::BSCall(double r_, double d_,
               double T_, double Spot_,
               double Strike_)
             :
               r(r_),d(d_),
               T(T_),Spot(Spot_),
               Strike(Strike_)
{}

double BSCall::operator()(double Vol) const
{
    return BlackScholesCall(Spot,Strike,r,d,Vol,T);
}
```

The constructor simply initializes the class data members, which are the parameters needed to price a call option under Black–Scholes except the volatility. The `operator()` takes in the volatility and then invokes the Black–Scholes formula.

This is the simplest possible implementation of the class. If we were truly worried about efficiency considerations, we could code the formula directly and precompute as much of it as possible in the constructor. We could have for example a class data member, `Moneyness`, set to the log of Spot divided by Strike, and then we would not have to compute it every time.

9.3 Bisecting with a template

In the previous section, we showed how we could define a class for which the syntax `f(x)` makes sense when `f` was an object of the class, and `x` was a `double`. We still need to get the object `f` into our solver routine, however. We do so via templatization. The basic idea of templatization is that you can write code that works for many classes simultaneously provided they are required to have certain operations defined with the same syntax. In this case, our requirement is that the class should have

```
double operator()( double ) const
```

defined, and thus that the syntax `f(y)` is well-defined for class objects as we discussed above.

We present the `Bisection` function in `Bisection.h`:

Listing 9.3 (`Bisection.h`)

```
template<class T>
double Bisection(double Target,
                 double Low,
                 double High,
                 double Tolerance,
                 T TheFunction)
{
    double x=0.5*(Low+High);
    double y=TheFunction(x);

    do
    {
        if (y < Target)
            Low = x;

        if (y > Target)
            High = x;

        x = 0.5*(Low+High);

        y = TheFunction(x);
    }
    while
        ( (fabs(y-Target) > Tolerance) );

    return x;
}
```

We only present a header file, since for template code we cannot precompile in a source file – we do not know the type of the object T. The function is quite simple. We specify that it is templatized via the `template<class T>` at the top. If we invoke the function with the template argument BSCall via `Bisection<BSCall>` then every T will be converted into a BSCall before the function is compiled. Once we have fixed the type of the template argument, there is really very little to the function. We take the midpoint of the interval evaluate the function there, and switch to the left or right side of the interval by redefining Low and High until the value at the midpoint is close enough to the target, and we then return.

Note that we have defined the type of the function object passed in as T The-Function: we could equally well have put const T& TheFunction. The syntax we have adopted involves copying the function object, and is therefore arguably less good than the alternative. The reason I have done it this way is to highlight the fact that the standard template library always uses the former syntax. A consequence of this is that one needs to be careful when using function objects with the STL not to define function objects which are expensive to copy (or, even worse, impossible to copy.)

We now give a simple example of an implied volatility function:

Listing 9.4 (SolveMain1.cpp)

```cpp
/*
Needs
    BlackScholesFormulas.cpp
    BSCallClass.cpp
    Normals.cpp
*/
#include <Bisection.h>
#include <cmath>
#include <iostream>
#include <BSCallClass.h>
#include <BlackScholesFormulas.h>

using namespace std;

int main()
{
    double Expiry;
    double Strike;
    double Spot;
    double r;
    double d;
    double Price;

    cout << "\nEnter expiry\n";
    cin >> Expiry;

    cout << "\nStrike\n";
```

```
cin >> Strike;
cout << "\nEnter spot\n";
cin >> Spot;

cout << "\nEnter price\n";
cin >> Price;

cout << "\nr\n";
cin >> r;

cout << "\nd\n";
cin >> d;

double low,high;

cout << "\nlower guess\n";
cin >> low;

cout << "\nhigh guess\n";
cin >> high;

double tolerance;

cout << "\nTolerance\n";
cin >> tolerance;

BSCall theCall(r,d,Expiry,Spot,Strike);

double vol =
     Bisection(Price,low,high,tolerance,theCall);
double PriceTwo =
     BlackScholesCall(Spot,Strike,r,d,vol,Expiry);

cout << "\n vol " << vol << " pricetwo "
                         << PriceTwo << "\n";

double tmp;
cin >> tmp;

return 0;
}
```

As usual, we input all the necessary parameters. We then put them together to create a BSCall object. We then call the Bisection function to find the volatility. Note that we have not put Bisection<BSCall>. The compiler deduces the type of the template argument from the fact that our final argument in the function call is theCall. We only have to specify the template argument when the compiler does not have a sufficient amount of other information to deduce it.

Our function finishes by getting the price via the Black–Scholes functions for the implied volatility that was found. If everything is working correctly then this will give the original price inputted.

9.4 Newton–Raphson and function template arguments

We now want to adapt the pattern we have presented to work for Newton–Raphson as well as for bisection. The fundamental difference from a design viewpoint is that Newton–Raphson involves two functions, the value and the derivative, whereas bisection involves just one. One solution would be simply to pass in two function objects: one for the value and another for the derivative. This is unappealing, however, in that we would then need to initialize a set of parameters for each object and we would have to be careful to make sure they are the same. More fundamentally, the value and the derivative are really two aspects of the same thing rather than two separate functions and so having two objects does not express well our conceptual model of them.

A second solution is to assume a name for the derivative function. After all, that is essentially what we did for the value function; it was a special name with special syntax but ultimately it was just assuming a name. Thus we could assume that the class had a method

```
double Derivative(double ) const
```

defined and at appropriate points in our function we would then put

```
TheFunction.Derivative(x).
```

This would certainly work. However, it is a little ugly and if our class already had a derivative defined under a different name, it would be annoying.

Fortunately, there is a way of specifying which class member function to call at compile time using templatization. The key to this is a pointer to a member function. A pointer to a member function is similar in syntax and idea to a function pointer, but it is restricted to methods of a single class. The difference in syntax is that the class name with a : : must be attached to the * when it is declared. Thus to declare a function pointer called Derivative which must point to a method of the class T, we have

```
double (T::*Derivative)(double) const
```

The function `Derivative` is a const member function which takes in a double as argument and outputs a double as return value. If we have an object of class T called `TheObject` and y is a double, then the function pointed to can be invoked by

```
TheObject.*Derivative(y)
```

Whilst the syntax is a little cumbersome, the key is to realise that it is the same for ordinary function pointers except that `T::` must be added to the `*` for declarations and `TheObject.` must be added for invocations.

We can now use a function pointer to specify both the derivative and the value. As we would like to avoid the time spent on evaluating the pointers, we can make them template parameters rather than arguments to our function. This means that the compiler can treat them just like any other function call, and as their types are decided at compile time they can be `inlined`.

Our Newton–Raphson routine is therefore as follows:

Listing 9.5 (`NewtonRaphson.h`)

```
template<class T, double (T::*Value)(double) const,
                  double (T::*Derivative)(double) const >
                  double NewtonRaphson(double Target,
                                        double Start,
                                        double Tolerance,
                                        const T& TheObject)
{
    double y = (TheObject.*Value)(Start);

    double x=Start;

    while ( fabs(y - Target) > Tolerance )
    {
        double d = (TheObject.*Derivative)(x);

        x+= (Target-y)/d;

        y = (TheObject.*Value)(x);
    }
    return x;
}
```

We have three template parameters: the class, the pointer to the value function for that class, and the pointer to the derivative function for that class. The routine is short and simple, now that we have the right structure. As usual, we keep repeating until close enough to the root. We have not included the checks required to ensure that the loop does not repeat endlessly if the sequence fails to converge to a root, but such additions are easily put in.

9.5 Using Newton–Raphson to do implied volatilities

Now that we have developed a Newton–Raphson routine, we want to use it to compute implied volatilities. Our class will therefore have to support pricing as a function of volatility and the vega as a function of volatility. As before, the other parameters will be class data members which are not inputted in the constructor rather than via these methods. We present a suitable class in BSCallTwo.h and BSCallTwo.cpp.

Listing 9.6 (BSCallTwo.h)

```
#ifndef BS_CALL_TWO_H
#define BS_CALL_TWO_H

class BSCallTwo
{

public:
    BSCallTwo(double r_, double d_,
                        double T, double Spot_,
                        double Strike_);

    double Price(double Vol) const;
    double Vega(double Vol) const;

private:
    double r;
    double d;
    double T;
    double Spot;
    double Strike;
};
#endif
```

The methods just call the relevant functions. As before, we could optimize by pre-computing as much as possible, and inlining the methods.

Listing 9.7 (BSCallTwo.cpp)

```cpp
#include <BSCallTwo.h>
#include <BlackScholesFormulas.h>


BSCallTwo::BSCallTwo(double r_, double d_,
                     double T_, double Spot_,
                     double Strike_)
                :
                r(r_),d(d_),
                T(T_),Spot(Spot_),
                Strike(Strike_)
{}

double BSCallTwo::Price(double Vol) const
{
    return BlackScholesCall(Spot,Strike,r,d,Vol,T);
}

double BSCallTwo::Vega(double Vol) const
{
    return BlackScholesCallVega(Spot,Strike,r,d,Vol,T);
}
```

We present an example of using the combination of `NewtonRaphson` and `BSCallTwo` in `SolveMain2.cpp`.

Listing 9.8 (SolveMain2.cpp)

```cpp
/*
Needs
    BlackScholesFormulas.cpp
    BSCallTwo.cpp
    Normals.cpp
*/
#include <NewtonRaphson.h>
#include <cmath>
#include <iostream>
#include <BSCallTwo.h>
#include <BlackScholesFormulas.h>
```

```
using namespace std;

int main()
{
    double Expiry;
    double Strike;
    double Spot;
    double r;
    double d;
    double Price;

    cout << "\nEnter expiry\n";
    cin >> Expiry;

    cout << "\nStrike\n";
    cin >> Strike;

    cout << "\nEnter spot\n";
    cin >> Spot;

    cout << "\nEnter price\n";
    cin >> Price;

    cout << "\nr\n";
    cin >> r;

    cout << "\nd\n";
    cin >> d;

    double start;

    cout << "\nstart guess\n";
    cin >> start;

    double tolerance;

    cout << "\nTolerance\n";
    cin >> tolerance;

    BSCallTwo theCall(r,d,Expiry,Spot,Strike);
```

```
    double vol=NewtonRaphson<BSCallTwo, &BSCallTwo::Price,
                    &BSCallTwo::Vega>(Price, start,
                                        tolerance, theCall);

    double PriceTwo =
        BlackScholesCall(Spot,Strike,r,d,vol,Expiry);

    cout << "\n vol " << vol << " \nprice two:" << PriceTwo
                                        << "\n";

    double tmp;
    cin >> tmp;

    return 0;
}
```

Our new main program is very similar to the one we had before. The main change is that this time we specify the template parameters for our NewtonRaphson function, whereas for the Bisection function we did not bother. The reason for the change is that there is not sufficient information for the compiler to deduce the types of the parameters. There is nothing to indicate which member functions of the class are to be used. Even for our class which only has two member functions, these two functions could equally well be the other way round as far the compiler knows.

9.6 The pros and cons of templatization

In this chapter, we have used template arguments to achieve re-usability whereas in other chapters, we have used virtual functions and polymorphism. There are advantages and disadvantages to each approach. The principal difference is that for templates argument types are decided at the time of compilation, whereas for virtual functions the type is not determined until runtime.

What consequences does this difference have? The first is speed. No time is spent on deciding which code to run when the code is actually running. In addition, the fact that the compiler knows which code will be run allows it to make extra optimizations which would be hard if not impossible when the decision is made at run time.

A second consequence is size. As the code is compiled for each template argument used separately, we have multiple copies of very similar code. For a simple routine such as a solver this is not really an issue but for a complicated routine, this could result in a very large executable. Another aspect of this is slower compiler times since much more code would have to be compiled. If we had several template parameters the size could multiply out of control. For example, suppose when designing our Monte Carlo path-dependent exotic pricer, we had templatized both the random number generator and the product. If we had six random number generators and ten products, and we wished to allow any combination then we would have sixty times as much code.

A third consequence is that it becomes harder for the user of the code to make choices. In the example of the exotics pricer, if the user was allowed to choose the number generator and the product via outside input, we would have to write code that branched into each of the sixty cases and within each branch called the engine and gathered results.

As well as the run time versus compile time decision, there are other issues with using templatized code. A simple one is that it is harder to debug. Some debuggers get confused by template code and, for example, refuse to set breakpoints within templates (or else set and ignore them.) Related to this is the fact that compilers will often not actually compile lines of template code that are not used. Thus if a templatized class has a method that is not called anywhere, the code will compile even if it has syntax errors. Only when a line is added that calls the particular method will the compiler errors appear. This can be infuriating as the error may show up a long time afterwards.

One way of avoiding these problems is first to write non-template code for a particular choice of the template parameter. This code can be thoroughly tested and debugged, and then afterwards the code can be rewritten by changing the particular parameter into a template parameter.

So when should one use templates and when use virtual functions? My preference is not to use templates unless certain conditions are met. These are that the routine should be short, and potentially re-usable in totally unrelated contexts. So for example, I would use templates for a numerical integration routine and a solver. I would also use templates for a container class; in fact I would generally use the templates given in the standard template library. I would not, however, use templates for an option pricing engine since the code would probably be long and is only relevant to a quite specific context.

The general trend in C++ is towards templates. The principal reason is that they are the way to achieve the same speed as lower-level languages. In general, languages exhibit a trade-off between abstraction and efficiency; C++ has always striven to achieve both. Templates are ultimately a way of achieving abstraction without sacrificing efficiency.

9.7 Key points

In this chapter we have looked at how to implement solvers using template code.

- Templates are an alternative to inheritance for coding without knowing an object's precise type.
- Template code can be faster as function calls are determined at compile time.
- Extensive use of template code can lead to very large executables.
- Pointers to member functions can be a useful way of obtaining generic behaviour.
- Implied volatility can only be computed numerically.

9.8 Exercises

Exercise 9.1 Modify the Newton–Raphson routine so that it does not endlessly loop if a root is not found.

Exercise 9.2 Take your favourite numerical integration routine, e.g. the trapezium rule, and write a template routine to carry it out.

Exercise 9.3 Write a routine to price a vanilla option by Monte Carlo or trees where the pay-off is passed in as a template parameter expressed via a function object.

10

The factory

10.1 The problem

Suppose we wish to design an interface which is a little more sophisticated than those we have used so far. The user will input the name of a pay-off and a strike, and the program will then price a vanilla option with that pay-off. We therefore need a conversion routine from strings and strikes to pay-offs. How might we write this?

One simple solution is to write a function that takes in the string and the strike, checks against all known types of pay-offs and when it comes across the right one, creates a pay-off of the right type. We would probably implement this via a switch statement. Our conversion routine would then have to include the header files for all possible forms of pay-off, and every time we added a new pay-off we would have to modify the switch statement. Clearly, this solution violates the open-closed principle as any addition involves modification.

In this chapter, we present a solution that allows us to add new pay-offs without changing any of the existing files. We simply add new files to the project. Our solution is a design pattern known as *the factory pattern*. It is so called because it can be used to manufacture objects. Whilst we restrict attention to a simple factory which manufactures pay-offs, the basic pattern can be used in much wider contexts.

10.2 The basic idea

Our solution requires each type of pay-off to tell the factory that it exists, and to give the factory a blueprint for its manufacture. In this context a blueprint means an identity string to distinguish that class and a pointer to a function that will create objects of that class.

How can we get the class to communicate with the factory, without explicitly calling anything from the main routine? The key lies in global variables. Every global variable is initialized when the program commences before anything else

157

happens. If we define a class in such a way that initializing a global variable of that class registers a pay-off class with the factory, then we have achieved what we wanted. This is possible because the initialization involves a call to a constructor, and we can make the constructor do whatever we want.

So for each pay-off class, we write an auxiliary class whose constructor registers the pay-off class with our factory, and we declare a global variable of the auxiliary class. In fact, as these auxiliary classes will all be very similar to each other, we adopt a template solution for defining these classes.

We also need a factory object for these auxiliary classes to talk to. We cannot make this factory object a global variable, as we have no control over the order in which the global variables are initialized. We need it to exist before the other globals are defined as they refer to it. Fortunately, there is a type of variable guaranteed to come into existence at the moment it is first referred to: the `static` variable.

Thus if the registration function contains a `static` variable which is the factory, then on the first call to the registration function the factory comes into existence. Recall that a `static` variable defined in a function persists from one call to the next, and only disappears when the program exits. So all the registration function calls will register the pay-off blueprints with the same factory.

However, we are not yet done because the creator function will need to have access to the same factory object as the registration function, and if the factory is hidden inside the registration function this will not be possible. The solution to this problem is known as the *singleton* pattern.

10.3 The singleton pattern

We saw in the last section that we need to define a factory via a `static` variable since it must come into existence as soon as it is referred to when registering the blueprints. We also saw that the same factory must be referred to by every registration, and that the same factory will be needed when creating the pay-offs from strings.

So what we need is a factory class for which an object exists as soon as it is required, and for this object to exist until the end of the program. We also do not want any other factory objects to exist as they will just confuse matters; everything must be registered with and built by the same factory.

The singleton pattern gives a way of creating a class with these properties. The first thing is that all constructors and assignment operators are made private. This means that factories can only be created from inside methods of the class this gives us firm control over the existence of factory objects. In order to get the one class object that we need, we define a very simple method that defines a class object as a static variable. Thus if our class is called `PayOffFactory`, we define a class method `Instance` as follows:

```
PayOffFactory& PayOffFactory::Instance()
{
    static PayOffFactory theFactory;
    return theFactory;
}
```

The first time that `Instance` is called, it creates the `static` data member `the-Factory`. As it is a member function, it can do this by using the private default constructor. Every subsequent time the `Instance` is called, the address of the already-existing `static` variable `theFactory` is returned. Thus `Instance` creates precisely one `PayOffFactory` object which can be accessed from anywhere by calling `PayOffFactory::Instance()`.

Note that `Instance` will have to be a `static` method of `PayOffFactory`, as the whole point is that it provides you with a `PayOffFactory` object, and it would be useless if you had to access it from an existing object. Note also that the meaning of `static` here for a *function* is quite different from the meaning above for a *variable*; for a function it means that the function can be called directly without any attachment to an object. One still has to prefix with the name of class, however.

We have now achieved what we needed: we have a way of creating a single factory which can be referenced from anywhere at any time in the program. Note that the name singleton pattern was chosen because precisely one object from the class exists.

10.4 Coding the factory

In the last section, we saw how the singleton pattern could be used to create a single factory accessible in any place at any time. We now use this pattern to implement the factory. As well as the instance method discussed above, we will need a method for registering pay-off classes and a method for creating then.

How will registration work? Upon registration, we need to know the string identifier for the specific pay-off class and the pointer to the function which actually creates the object in question. These will therefore be the arguments for the registration method. The factory will need to store this information for when the create pay-off method is called. This will require a container class.

Fortunately, there is a container in the standard template library which is designed for associating identifiers to objects. This container is called the `map` class. We therefore need a data member which is a map with template arguments `std::string` and pointers to create functions.

Finally, we need a method which turns a string plus a strike into a `PayOff` object. Our header file is therefore as follows:

Listing 10.1 (`PayOffFactory.h`)

```
#ifndef PAYOFF_FACTORY_H
#define PAYOFF_FACTORY_H
#include <PayOff3.h>

#if defined(_MSC_VER)
#pragma warning( disable : 4786)
#endif

#include <map>
#include <string>

class PayOffFactory
{
public:
    typedef PayOff* (*CreatePayOffFunction)(double );

    static PayOffFactory& Instance();
    void RegisterPayOff(std::string, CreatePayOffFunction);
    PayOff* CreatePayOff(std::string PayOffId,double Strike);
    ~PayOffFactory(){};

private:
    std::map<std::string, CreatePayOffFunction>
                TheCreatorFunctions;
    PayOffFactory(){}
    PayOffFactory(const PayOffFactory&){}
    PayOffFactory& operator=
                (const PayOffFactory&){ return *this;}
};
#endif
```

Note the `typedef`: this allows us to refer to pointers to functions which take in a double and spit out a `PayOff*` as `CreatePayOffFunction`. Without this `typedef`, the syntax would quickly become unmanageable. Note also that we make the `CreatePayOff` method return a pointer to a `PayOff`. The reason for this is that it allows the possibility of returning a null pointer if the identity string was not found; otherwise we would have to `throw` an error or return a default sort of pay-off.

We present the source code in `PayOffFactory.cpp`:

Listing 10.2 (PayOffFactory.cpp)

```cpp
#if defined(_MSC_VER)
#pragma warning( disable : 4786)
#endif


#include <PayOffFactory.h>
#include <iostream>
using namespace std;

void PayOffFactory::RegisterPayOff(string PayOffId,
                         CreatePayOffFunction CreatorFunction)
{
  TheCreatorFunctions.insert(pair<string,CreatePayOffFunction>
                         (PayOffId,CreatorFunction));
}


PayOff* PayOffFactory::CreatePayOff(string PayOffId,
                                    double Strike)
{
  map<string, CreatePayOffFunction>::const_iterator
                     i = TheCreatorFunctions.find(PayOffId);

  if  (i == TheCreatorFunctions.end())
  {
      std::cout << PayOffId
      << " is an unknown payoff" << std::endl;
      return NULL;
  }

  return (i->second)(Strike);
}


PayOffFactory& PayOffFactory::Instance()
{
   static PayOffFactory theFactory;
   return theFactory;
}
```

Other than for the Instance() method, which we have already discussed, the methods are really just wrappers for the inner map object.

In case the reader is not familiar with the STL map container, we discuss a little how it works. A map is a collection of pairs. We used the pair class to store the nodes of a tree in Chapter 8. Recall that a `pair` is a simple class consisting of two public data members known as `first` and `second`. The types of these data members are template parameters. When working with a map, `first` is the key or identifier used to look up the object we wish to find which is stored in `second`.

For us this means that the type of the map is

```
map<std::string, CreatePayOffFunction>
```

and every `pair` that we use will be of the same type. The `insert` method is used to place `pairs` of `strings` and `CreatePayOffFunctions` into the map. A `map` has the property that each key is unique so if you insert two `pairs` with the same key then the second one is ignored. For us, this means that if we give two `PayOff` classes the same string identifier only one will be registered. It is possible to examine the return type of the `insert` to determine whether the insertion was successful. The method `RegisterPayOff` carries out this insertion for our factory.

The retrieval is carried out in `CreatePayOff`: a `string` is passed in and the `find` method of `map` is used. This method returns a `const_iterator` pointing to the `pair` which has the correct key (i.e. `first` element) if such a `pair` exists, and otherwise the iterator points to the end of the map. For the reader who is not familiar with iterators, an iterator is an abstraction of a pointer and works in similar fashion. Just like pointers, they can be dereferenced via `*` or `->`. A `const_iterator` is similar to a non-`const` pointer to `const` objects. That is, the iterator's value can be changed, but the value of the thing it points to cannot.

Our method therefore uses `find` to get an iterator. This iterator is then checked to see if the look-up succeeded. If it failed we print an error message and return a null pointer. If is succeeded, we take the `second` element of the `pair` pointed to, which is a function pointer, dereference it and call it with argument `Strike`. Since this function's job is to create a `PayOff` object of the relevant type and return a pointer to it, we have achieved our objective; objects of any class previously registered can be created by entering the appropriate string and strike.

Whilst we certainly could have programmed our factory without using the STL `map` class, its existence certainly made the task much easier. We refer the reader to Stroustrup, [31], Section 17.4, Josuttis, [12], and Meyers, [20], for further information on the `map` class.

10.5 Automatic registration

We discussed above how we could manage registration of `PayOff` classes by using global variables. Here we look at how to carry out this out. As we mentioned above, the code is the same except for class names for each registration so it makes sense to use a template class. We present this template code in `PayOffConstructible.h`:

Listing 10.3 (PayOffConstructible.h)

```
#ifndef PAYOFF_CONSTRUCTIBLE_H
#define PAYOFF_CONSTRUCTIBLE_H

#if defined(_MSC_VER)
#pragma warning( disable : 4786)
#endif

#include <iostream>
#include <PayOff3.h>
#include <PayOffFactory.h>
#include <string>

template <class T>
class PayOffHelper
{
public:
    PayOffHelper(std::string);
    static PayOff* Create(double);
};

template <class T>
PayOff* PayOffHelper<T>::Create(double Strike)
{
    return new T(Strike);
}

template <class T>
PayOffHelper<T>::PayOffHelper(std::string id)
{
    PayOffFactory& thePayOffFactory = PayOffFactory::Instance();
    thePayOffFactory.RegisterPayOff(id,PayOffHelper<T>::Create);
}
#endif
```

The helper class we define here has to do two things. It must define a constructor that carries out the registration of the class defined by the template parameters, and it must define a function which will carry out the creation so we have something to use in the registration process!

The constructor takes in a `string` as an argument; this string will be needed to identify the class being registered. The constructor simply first calls `Instance` to get the address of the factory object, and then calls the `RegisterPayOff` method of the factory to carry out the registration. Note that the constructor does not actually do anything as regards the actual object being created! In fact, the class has no data members so it would not be possible to do anything.

The method `Create` defines the function used to create the pay-off object on demand. Note that it is `static` as it should not be associated to any particular class object. The function simply calls the constructor for objects of type T with argument `Strike`. Of course, there is something slightly subtle here in that the specification of the template parameter, T, is making the choice of which object to construct. Note that we use `new` as we want the created object to persist after the function is finished. One consequence of this is that the object will have to be properly `deleted` at some point.

`PayOffRegistration.cpp` includes an example of using the `PayOff Helper` class.

Listing 10.4 (`PayOffRegistration.cpp`)

```
#include <PayOffConstructible.h>

namespace
{
  PayOffHelper<PayOffCall> RegisterCall("call");

  PayOffHelper<PayOffPut> RegisterPut("put");
}
```

Note that if we were defining a new class, we would probably put this registration in the source file for the class but as we have already defined the call and put classes, we do not do so here. The registration file is quite short. We define two global variables, `RegisterCall` and `RegisterPut`. These are of type `PayOffHelper<Call>` and `PayOffHelper<Put>`. As global variables, they are initialized at the start of the program. This initialization carries out the registration as required. Note that we have put a `namespace` command around the declaration. This means that the variables are in an unnamed namespace and as such are invisible to the rest of the program. So the variables are both global and invisible. Why do we want them invisible? Their purpose is purely to perform the registration, and once that has been done we have no further use for them so it is best to put them out of sight and out of temptation's reach.

10.6 Using the factory

Now we have done all the set-up work, how do we use the factory? We give a very simple example in `PayFactoryMain.cpp`.

Listing 10.5 (`PayFactoryMain.cpp`)

```
/*
Uses
    PayOff3.cpp
    PayOffBridge.cpp
    PayOffFactory.cpp
    PayOffRegistration.cpp
*/


#include <PayOff3.h>
#include <PayOffConstructible.h>
#include <PayOffBridge.h>
#include <PayOffFactory.h>
#include <string>
#include <iostream>
using namespace std;

int main()
{
    double Strike;
    std::string name;

    cout << "Enter strike\n";
    cin >> Strike;

    cout << "\npay-off name\n";
    cin >> name;

    PayOff* PayOffPtr =
        PayOffFactory::Instance().CreatePayOff(name,Strike);

    if (PayOffPtr != NULL)
    {
        double Spot;

        cout << "\nspot\n";
        cin >> Spot;
```

```
        cout << "\n" << PayOffPtr->operator ()(Spot) << "\n";
        delete PayOffPtr;
    }

    double tmp;
    cin >> tmp;
    return 0;
}
```

This routine is very simple but illustrates the important points. The user inputs spot, strike and the name of the option. If an option with that name has been registered then the pay-off is computed, and the object is then deleted.

The important point here is that the name definitions are carried out in the file PayOffRegistration.cpp, and this file is not seen directly by any of the other files including the main routine. If we wanted to add another PayOff, say the forward, we could so without modifying any of the existing files. In fact, all we would have to do is add the header and source file for the forward, and in a new file PayOffForwardRegistration.cpp add the declaration

```
PayOffHelper<PayOffForward> RegisterForward("forward");
```

As we originally required, this would not require recompilation of any of the original files. We have therefore achieved our original objective of an open-closed pattern.

10.7 Key points

In this chapter we have developed the factory pattern and the singleton pattern in order to give a method of adding new pay-off classes to an interface without modifying existing files.

- The singleton pattern allows us to create a unique global object from a class and provide a way of accessing it.
- The factory pattern allows us to add extra inherited classes to be accessed from an interface without changing any existing files.
- The factory pattern can be implemented using the singleton pattern.
- The standard template library map class is a convenient way to associate objects with string identifiers.
- Placing objects in an unnamed namespace is a way of ensuring that they are not accessed elsewhere.
- We can achieve automatic registration of classes by making their registration a side-effect of the creation of global variables from a helper class.

We will return to the factory pattern in Chapter 14; there we will see how to implement it in a generic way so that one implementation will do forever.

10.8 Exercises

Exercise 10.1 Write a straddle class and register it with the factory.

Exercise 10.2 Our class cannot handle a double digital as it needs two strikes. Work out a solution that will handle options with multiple parameters.

Exercise 10.3 Integrate the factory with a Monte Carlo routine.

Design patterns revisited

11.1 Introduction

In this chapter, we revisit and catalogue the design patterns from earlier chapters. We also mention a few other patterns we have not studied which the reader may find helpful. Finally, we discuss further reading on the topic of design patterns.

The design patterns we have studied are a small subset of those in the classic book on the topic: *Design Patterns*, [7], which is often referred to as the 'Gang of Four' book. As well as listing the patterns, the authors attempt to classify the patterns according to the contexts in which they are used. In this final chapter, we revisit the patterns the we have studied in the context of that classification. We also mention some patterns discussed there which we have not examined here.

11.2 Creational patterns

A creational pattern is a pattern that deals primarily with the creation of new objects. Their purpose is to abstract the creation process which helps the system to be developed independently of the types of individual objects. In fact, sometimes all we can be sure of about these objects, is what class they are inherited from, or in other terms what interface they implement.

11.2.1 Virtual copy constructor

We have extensively used the concept of cloning. We need a copy of an object, we do not know its type so we cannot use the copy constructor so we ask the object to provide a copy of itself. Note the general philosophy here is that the object knows more about itself than we do so we ask it to help us out. Note we could easily modify this pattern to ask the object to make a default object from its class, as once again it knows its class type and we do not. In [7] virtual constructors are known as the 'Factory Method'.

11.2.2 The factory

This is called the 'abstract factory' in [7]. The purpose of this pattern is to allow us to have an object that spits out objects as and when we need them. This means in particular that responsibility for creating objects of the relevant type lies with a single object. We thus gain greater control over their creation, which yields greater flexibility in changing the objects used. We principally used this pattern to give an easily extended interface. In particular, we saw that the pattern allows the addition of new classes to an interface without the rewriting of any code.

11.2.3 Singleton

We used the singleton pattern to implement our factory. The big advantage of the singleton pattern is that there is a single copy of the object which is accessible from everywhere, without introducing global variables and all the difficulties they imply. Note that if for some reason we wanted more than one copy of the object to exist then we could easily modify the pattern to given us a doubleton or a tripleton and so on. For example, we could define more than one method that performed similarly to our `Instance` method.

11.2.4 Monostate

We have not examined the monostate pattern nor is it covered in *Design Patterns*, however, it is a useful alternative to the singleton pattern. Rather than only allowing one object from the class to exist, we allow an unlimited number but make them all have the same member variables. Thus all the objects from the class act as one. The way we do this is by making all the data members `static`. This approach allows us to treat each object from the class like any other, although they are all really the same object. Our factory could easily have been implemented using this pattern.

11.3 Structural patterns

A structural pattern is one that deals mainly with how classes are composed to define more intricate structures. They allow us to design code that has extra functionality without having to rewrite existing code.

11.3.1 Adapter

The adapter is a class that translates an interface into a form that other classes expect. It is most useful when we wish to fit code into a structure for which it was not originally designed, either because we have changed our way of doing things or because the code originates elsewhere. For example, if we download a library from the web, its interface is unlikely to conform to what we have been using. By adapting the interface, we can seamlessly integrate it into existing code. We gave an example of the adapter when implementing random number generators.

11.3.2 Bridge

The bridge is similar to the adapter in that it defines an interface, and acts as an intermediary between a client class and the classes implementing the interface. Thus the implementing class can easily be changed without the client class being aware of the change. The main difference between the bridge and the adapter is that the bridge is intended to define an intermediary interface from the start, whereas the adapter is introduced a later stage in order to solve incompatibilities. We used the bridge to create the `PayOff` and `Parameters` objects.

11.3.3 Decorator

The decorator patterns allows us to change the behaviour of a class at run-time without changing its interface. We add a wrapper class that processes incoming or outgoing messages and then passes them on. We saw that a decorator could be used to implement anti-thetic sampling when studying random number generation. We also saw that it could be used to create convergence tables of statistics when designing statistics gatherers. An attractive aspect of decoration is that we can decorate as many times as we like since the interface after decoration is the same as the interface before.

11.4 Behavioural patterns

Behavioural patterns are used for the implementation of algorithms. They allow us to vary aspects of algorithms interchangeably. They can also allow us to re-use algorithms in wildly unrelated contexts.

11.4.1 Strategy

In the strategy pattern, we defer an important part of algorithm to an inputted object. This allows us to easily change how this particular part of the algorithm behaves. We have used this pattern implicitly and explicitly all through the book. By making the `PayOff` or `VanillaOption` an input to our pricer, we are using this pattern. Less trivially, we made the random number generator an input to our Monte Carlo, and the generator is a key part of the algorithm.

11.4.2 Template

In the template pattern, rather than inputting an aspect of the algorithm, we defer part of the algorithm's implementation to an inherited class. The base class thus provides the structure of how the different parts of the algorithm fit together, but does not specify all the details of the implementation. We adopted this approach when designing our exotics Monte Carlo pricer; there we defined the process for the stock price evolution in an inherited class. This allowed the possibility of pricing exotics using a different model in the future.

11.4.3 Iterator

We have only briefly mentioned iterators; however, they are an important component of the standard template library. An iterator is essentially an abstraction of a pointer. As such it should be possible to dereference it, i.e. look at what it points to, increment and decrement it. The idea of iterators is that one can be defined for any sort of data structure, and so if an algorithm is defined in terms of iterators it can be applied to any sort of data structure. In the STL, algorithms take the type of the iterator as a template argument, which allows this generality to be implemented.

11.5 Why design patterns?

Why think in terms of design patterns? What has this classification bought us? The first simple thing is that it becomes much easier to explain our code to someone else; remember that re-use is ultimately defined socially not analytically. When describing our code to someone else, if we can describe a class by saying this is such and such standard pattern then they immediately have a mental model of how it works from their previous familiarity with that pattern.

A second advantage is that by having familarity with a collection of standard design patterns, we gain an immediate toolbox for solving any problem put in front of us. Thus when confronted with a programming problem, we can approach it by thinking "what design pattern is appropriate here?" rather than by attempting to solve it from scratch. Even if none of the known patterns are appropriate, examination of the problem through the lens of design patterns will help us to solve it. In particular, the knowledge of why the patterns are inappropriate will aid us in developing a solution.

Of course, most programmers have patterns they implicitly use regularly. Indeed, many experienced programmers reading this book may feel they are only learning a formalization of what they did in any case. The advantage for them in using patterns is that it will help them to think more clearly about what they do and why.

11.6 Further reading

There are now many books on the topic of design patterns. We mention a few that the author has found useful. One good and straightforward book which was deliberately written as an easier companion to *Design Patterns* is *Design Patterns Explained* by Shalloway & Trott. The authors carefully go through many patterns explaining in simple language the concepts introduced in the original book.

C++ Programming: with Design Patterns Revealed by Muldner is another introductory book which is accessible, and takes the point of view that C++ should be learnt from the start in terms of design patterns.

Modern C++ Design by Alexandrescu is a more advanced book. It covers many more intricate ideas using templates than we have had the opportunity to cover here.

As well as books explicitly on design patterns, a C++ programmer needs many other standard texts. Some favourites of the author are

Effective C++, *More Effective C++* and *Effective STL* by Scott Meyers. These books are collections of programming gems by a C++ expert.

The C++ Programming Language by Bjarne Stroustrup. This is the ultimate reference book on C++ by the man who invented the language.

The C++ Standard Library by Nicolai Josuttis. This is a comprehensive description of the standard library that ships with any C++ compiler.

My favourite introductory book on object-oriented programming is

The Tao of Objects by Gary Entsminger. It's an easy read and concentrates on introducing the basic ideas of OO design.

There are now a number of books on C++ and numerical techniques in finance. Of all of these, the one closest in style and approach to this book is the forthcoming:

Quantitative Finance: An Object-oriented Approach + C++ by Eric Schlögl.

The books by Daniel Duffy are also worthwhile but take a different tack with more emphasis on templates and less on virtual functions.

11.7 Key points

- Design patterns can be classified into behavioural, structural and creational patterns.

- Behavioural patterns are used for the implementation of algorithms.
- Structural patterns deal with how classes are composed to create more intricate designs.
- A creational pattern deals primarily with the creation of new objects.

11.8 Exercise

Exercise 11.1 Implement the factory from Chapter 10 using the monostate pattern instead of the singleton pattern.

12

The situation in 2007

12.1 Introduction

The first eleven chapters of this book were written in the summer of 2002. Inevitably, both C++ and quantitative finance have moved on in the last five years, and, in addition, my view of the two subjects has evolved. In this brief chapter, I want to discuss some of the changes and set the stage for the newly added chapters.

12.2 Compilers and the standard library

In 2002, the most popular compiler for C++ was Visual Studio 6.0. Other popular compilers were g++ 2.95 and Borland 5.5. I therefore targeted the book and the code at those three compilers. Today, Visual Studio has gone through a couple of versions and the most popular version is 8.0. In addition, the upgrade process has been faster than usual in that Microsoft decided to make the "Express" version free. This version contains the full optimizing compiler and IDE (integrated development environment) but omits various added features which are not particularly important to the lone developer. The open source compiler g++ has also evolved and the most recently released version is as part of gcc 4.2.0. In addition, the standard libraries that ship with the compilers have been updated.

What difference does this make? The biggest difference for the programmer is that the compilers and libraries are much closer to the C++ ANSI/ISO standard, which was ratified in 1997. They are still not fully compatible in that they do not implement the `export` keyword. In fact, the only compiler that does is the Comeau Compiler (www.comeaucomputing.com), which is the only fully compatible compiler. If you are wondering what the `export` keyword is; don't. (If you must know, it gives an alternate way of implementing templates that does not require all the code to be in the header file.)

Most of the advantages relate to complicated template code, which we do not attempt to discuss in this book, but refer the reader to [35] for how to make a Turing machine run at **compile time** using template code.

However, some of the advantages are more mundane and are definitely worth mentioning. One example is that `for` loops `variables` are now properly scoped. If you declare

```
for (int i =0; i < 10; ++i)
{
 // do stuff
}


i =0;
```

then you will get a compilation error unless you have declared `i` before the `for` loop; this was not true in 6.0. On the other hand, the following code did not compile in 6.0

```
for (int i =0; i < 10; ++i)
{
 // do stuff
}


for (int i =0; i < 10; ++i)
{
 // do more stuff
}
```

as the variable `i` is declared twice in the same scope. This can be quite annoying if you are writing for cross-platform compatibility! One solution is to put {,} around the first `for` loop forcing the correct behaviour in 6.0 without affecting behaviour on the other compilers.

Another relevant improvement is that changing the return type of inherited class pointers is supported in 8.0. So we can now make our `clone` method return a pointer of an inherited class type. That is we can code

```
class Base
{
public:
    virtual Base* clone() const=0;

// etc

};
```

```
class Inherited : public Base
{
public:
    virtual Inherited* clone() const;

// etc

};
```

and not get an error about the fact that the return type has changed. This is mainly useful if we have 3 level hierarchies. For example, suppose we decided to implement pay-off type via inheritance (I don't recommend this, but it serves to illustrate the point), so we have a base class EquityOption, we inherit VanillaOption from this, and CallOption from VanillaOption.

Now suppose we declare clone in EquityOption, it will return a pointer of type EquityOption*. Under the old rules, the clone method of CallOption would have had to return a pointer of type EquityOption. This would have made it impossible to use Wrapper<VanillaOption>, which would be rather inconvenient.

Under the new rules, we return a pointer of the most derived type and it can always be treated as a pointer to a class further up the hierarchy so we have no problems.

Another big change that is very convenient is that the standard template library has range-checking in Debug mode in Visual C++ 8.0. With previous versions of the library, this was always a big disadvantage when working with the STL vector in that one either used .at all the time and suffered a performance penalty in Release mode, or spent ages trying to track down out-of-range errors which were not reported. Note that if you are using another compiler, you can get an alternative implementation of the standard library from www.stlport.org which does feature range-checking.

12.3 Boost

The Boost project is an open source library designed to extend the C++ standard library. It can be found at www.boost.org. The code is heavily peer-reviewed and required to work across multiple compilers. The intention is that the libraries incorporated in Boost will become part of the C++ Standard in the future, and, indeed, it is already planned to incorporate many of them. Because of this, the licence is very unrestrictive and allows the user basically to do whatever they want with the code. This is different from the GNU licence, which restricts use of the code to applications that distribute source code and allow the user to do the same. From the Boost website, here are some of the requirements met by the licence:

- Must grant permission without fee to copy, use and modify the software for any use (commercial and non-commercial).
- Must require that the license appear with all copies [including redistributions] of the software source code.
- Must not require that the license appear with executables or other binary uses of the library.
- Must not require that the source code be available for execution or other binary uses of the library.

What all this means is that you can use code from Boost in your work without ever having to worry about licensing issues.

The main downside of Boost used to be that the installation process was annoying, they had their own customized routines for installation that you had to use. However, there is now an installer that gives you the pre-compiled binaries for the libraries if you use Visual C++ 7.1 or Visual C++ 8.0.

There are far too many libraries to attempt to discuss them here. However, two of particular interest to quantitative analysts are the random number library, and the multi-dimensional array library. We will discuss the smart pointers library a little in Chapter 13. Books are now being written on Boost and a useful one is *Beyond the C++ Standard Library* by Karlsson. Boost is used heavily by the Quantlib project.

12.4 QuantLib

QuantLib is the biggest and most successful open-source project for quantitative finance. It is a large repository of C++ code and can be found at quantlib.org. Similarly to Boost, the license is very unrestrictive, allowing free use in commercial software. The objective is to provide a large pricing library for derivatives that can used for many purposes. The code is very much structured as one library, and whilst individual routines can be cut out, it is not designed to facilitate this. Familiarity with QuantLib will certainly be an important skill in the future. The author of this book is now a developer on the project.

As well as providing a C++ library, QuantLib comes with code for building interfaces to various applications, particularly EXCEL. The main difficulty with QuantLib is that requires a certain amount of sophistication to make sense of the library, so once you are comfortable with C++ start learning, but wait until you are.

12.5 xlw

Although in the first part of this book, we concentrated on writing console applications, it is actually rare for quants to work in that way. One of the most common modes of working is to write EXCEL plug-ins, known as xlls. These add extra functions to EXCEL, which can be called from the spread-sheet. These work via the C API, which was never very well documented. In order to make interfacing easier, Jerome Lecomte wrote a C++ wrapper called xlw. This made the whole process much simpler but still involved writing a lot of repetitive code. Ferdinando Ametrano took over the project and ported it to `sourceforge.net`. The author of this book then took over the project and made various changes. The biggest of these is that the interfacing code is now automatically generated; this means that the user has to do very little. Another important aspect of the package is that it works with the free MingW g++ compiler as well as Visual C++ 6.0, 7.1, and 8.0. We will discuss using examples from xlw in the rest of the book, particularly in Chapter 14 where we develop a generic factory using the implementation in xlw as an example. We discuss how to use the xlw project in detail in Chapter 15.

The code for xlw can be obtained from `xlw.sourceforge.net`.

12.6 Key points

- The most popular compilers are now Visual Studio 8.0 and gcc 4.2.0.
- The new compilers are closer to being compliant with the C++ standard.
- `Boost` is a high-quality free library of C++ code.
- QuantLib is the largest library of open source C++ code for quantitative finance.
- Most work in quantitative finance is done via interfacing with EXCEL.
- xlw provides an easy way to interface with EXCEL.

12.7 Exercises

Exercise 12.1 Install `Boost` on your computer.

Exercise 12.2 Interface the `Boost` random number classes with the path-dependent exotic option pricer developed here.

Exercise 12.3 Download and build Quantlib!

13

Exceptions

13.1 Introduction

Up till now we have focussed on clarity and code reusability, we have not considered how to cope with things going wrong at run time. The mechanism in C++ designed for coping with errors is throwing an exception. Writing code that functions well in the presence of exceptions raises a host of issues that did not exist before. We will look at some of these and see how most of them can be avoided by following some simple rules.

Exceptions are raised by the `throw` command. We specify as an argument an object, X, of any type Y. Execution then immediately moves to the end of the current scope and objects going out of scope are destroyed. If there is a `catch` command at the end of the scope, which catches objects of type Y, then control passes to the scope of the `catch` command. If not, then control passes to the end of the enclosing scope, and this keeps happening until the exception is caught, or the enclosing scope is the end of the program and execution terminates, i.e. your program crashes. Note that we can always do a catch-all statement with `catch(...)`.

The great virtue of this approach is that we do not have to test the return value of every function or method call to ensure that the last call did not generate an error. The great downside is that code execution order becomes a lot less predictable, and this can cause problems. In particular, if we write code for cleaning up at the end of a scope it may be bypassed by a `throw`, resulting in things being left in a poor state.

This is particularly a problem when memory allocation is in use. Consider the following code snippet with the `PayOff` class as in Chapter 4.

```
double evaluate(const PayOff& p,
                double y)
{
  PayOff* payOffPtr = p.clone();
```

179

```
    double x = (*payOffPtr)(y);
    delete payOffPtr;
    return x;
}
```

This is a little artificial in that no useful purpose is served by making a copy, but if operator() were non-const, this could be useful. In any case, it serves to illustrate a point: it is possible that calling operator() will throw an exception. This will be caught somewhere outside the function. The catcher will have no idea that payOffPtr needs to be deleted. The effect will be that the memory allocated by the clone will never be deallocated. If this happens enough times, your application will run out of memory and crash. If we are writing a stand-alone program that does not bother to catch exceptions, this is not such an issue but as soon as we are working in a system which is not supposed to die every time an exception is thrown, it is a real problem.

Given that any piece of code we call may throw an exception, to be sure that our code is correct and remains correct (for the code called may change its implementation), we are forced to program defensively as if an exception could be thrown at any time.

13.2 Safety guarantees

There are two standard safety guarantees:

- The *weak* guarantee: the object and program are left in a valid state, and no resources have been leaked.
- The *strong* guarantee: if an exception is thrown during an operation (e.g. a call to a method or a function), then the program is left in the state it was at entry to the operation.

The essential difference here is that with the weak guarantee an object's state can change even though the operation failed, whereas with the strong guarantee the class is promising to undo all changes before throwing.

Clearly, the strong guarantee is harder to implement than the weak one. However, it is important to realize that code that is not written with exception safety in mind will satisfy neither. The weak guarantee is also sometimes called the *basic* guarantee.

13.3 The use of smart pointers

Consider again the example of the introduction. What we want to happen is that when the function is exited the memory allocated by the clone command is deallocated by a call to delete. Exiting can occur either in the conventional way via the return statement, or by the exception being thrown. For both of these, all automatic (i.e. ordinary local) variables are destroyed at the end of the scope. So the solution is to make the deletion a side-effect of these destructions.

We have already looked at one smart pointer Wrapper<T>. If we use it here, the code snippet becomes

```
double evaluate(const PayOff& p,
                double y)
{

    Wrapper<PayOff> payOffPtr(p);
    double x = (*payOffPtr)(y);
    return x;
}
```

Recall that the Wrapper class will call the clone method internally. The delete command is no longer necessary because the destructor of Wrapper calls it automatically.

As written, Wrapper<T> cannot be used to take ownership of a raw pointer since it has no constructors that take pointers. However, we can easily add to the file Wrapper.h (see Listing 5.6) an extra constructor in the public section of the class

```
Wrapper(T* DataPtr_ )
{
    DataPtr =DataPtr_;
}
```

and then it would be legitimate to code

```
double evaluate(const PayOff& p,
                double y)
{
    PayOff* payPtr1 = p.clone();
    Wrapper<PayOff> payOffPtr(payPtr1);
    double x = (*payOffPtr)(y);
    return x;
}
```

and retain the automatic deletion of the allocated memory.

The Wrapper<T> class is just one example of a smart pointer. There are many examples both in Boost and the standard library. These generally vary according to what happens on copying the pointer.

There are four obvious solutions to copying:

(1) Copy the pointed-to object.
(2) Make copying illegal.
(3) Have the pointers share ownership of the object.
(4) Transfer ownership of the pointer to the new object.

The first of these is the approach adopted by the `Wrapper<T>` class. The main downsides of this approach are that copying may be slow and that it relies heavily on the writer of the pointed-to class having provided a `clone()` method.

With the second, we make the copy constructor and the assignment operator of the object `private`. Whenever a coder attempts to copy (or assign) the smart pointer, an error message saying that the copy constructor is not available is generated and the user is forced to find an alternate approach. If you want this sort of pointer, use the `scoped_ptr` class from Boost defined in `boost/scoped_ptr.hpp`. Note that the error is generated at compile time rather than run time, since it arises from access permissions to class methods – these are checked only at compile time.

An alternate implementation would be to make the copy constructor and assignment operator `throw`. This would be less desirable, however, in that the error would only be generated whilst the code was running, and if the code was rarely used, might take a long time to show up.

The third of these approaches is adopted by the boost shared pointer class: `shared_ptr` defined in `boost/shared_ptr.hpp` . We then essentially have a reference-counted pointer class (cf. Exercise 5.6). Every time the `shared_ptr` is copied, a count of how many pointers there are to the object is increased by one, and every time one is destroyed the count is decreased by one. When the count hits zero the pointed-to object is destroyed. We again do not have to worry about exception safety since when the last `shared_ptr` goes out of scope the object is deleted.

The main downside of `shared_ptr` is that there is only ever one copy of the object. This means that if one piece of code changes the object, then the object pointed to by all the copied pointers also changes since it is the same object. This means that the programmer has to think a little more than with the `scoped_ptr` or the `Wrapper` since you have linkage between not obviously connected things.

The last of these alternatives is used by the `auto_ptr` class in the standard library defined in the file `memory`. As with all the other smart pointers, when it goes out of scope the pointed-to object is deleted. However, suppose we code the following

```
double evaluate(const PayOff& p,
                double y)
{
    std::auto_ptr<PayOff> payPtr1 = p.clone();
```

```
    double z = (*payPtr1)(y);
    std::auto_ptr<PayOff> payPtr2(payPtr1);
    double x = (*payPtr1)(y);
    return x+z;
}
```

We will get a nasty run-time crash at the line where x is declared. Why? The copying of the object payPtr1 into payPtr2 *changes* the object payPtr1. This is very counter intuitive – we ordinarily expect copying an object to have no effect on the original object, but with auto_ptrs a great deal changes. All ownership is transferred to the new object and the first pointer becomes a null pointer.

This behaviour is occasionally useful when creating an object in a function or method and wishing to return a pointer to it. We want the client to take ownership of the object immediately without further copying and without having to use an unsafe raw pointer, and auto_ptr provides this facility. However, so does shared_ptr without the strange behaviour.

Ultimately, which smart pointer to use is a matter of personal style. The important thing is always to use one and to stay away from raw pointers. I personally almost always use the Wrapper, and very occasionally use shared_ptr and auto_ptr. The reason for preferring Wrapper is that it requires least thought: all deletions occur naturally, and all objects have intuitive copying behaviour. Its only real downside is that it makes object copying slower, but in numerical code you should avoid all copying within tight loops in any case, so this actually has little impact.

13.4 The rule of almost zero

The use of smart pointers brings us to a rule of programming. We previously studied the "rule of three"; this said that if you define one of copy constructor, assignment operator, and destructor for a class, then you should define all three. The "rule of almost zero" does not contradict this rule but supersedes it by saying that you should always be in the case of not defining any of them.

How do we avoid the shallow copy problem discussed in Chapter 4? We use smart pointers to ensure that a shallow copy is sufficient. Every data member will be either an ordinary object which can be copied, or a smart pointer which is copied and assigned in the fashion we have chosen. So if we want objects to be shared between copies, we use shared_ptr; if we want to make copying illegal, we use scoped_ptr; and if we want the pointed-to objects to be cloned, we use Wrapper.

There will be no memory leak issues because the smart pointers delete the pointed-to objects when the compiler-generated destructor is called. We do not waste time writing copy constructors or assignment operators, and we do not have to remember to update them when we change the data members of the class – forgetting to keep them in line with the class data members is a common source of bugs.

Why have we named it the "rule of almost zero" instead of the "rule of zero"? There is one case in which we must declare a destructor but only an empty one! Every time we have a class with abstract methods it is likely to be deleted via pointers to the base class and so we must declare a virtual destructor as discussed in Section 3.5.

13.5 Commands to never use

Some commands to never use are:

```
malloc
free
delete
new []
delete []
```

The first two of these are C commands not C++, and have been superseded by the versions of new and delete. You will get bizarre effects (i.e. crashes) if you try to mix the two sets of commands.

The delete command is never necessary because of smart pointers. As long as you ensure that anything created by new is owned by a smart pointer, you need never code delete. The only time I therefore use the delete command is when I am writing a smart pointer. However, with the advent of Boost, you should never need to do this – if a pointer doing what you want exists there, use that instead.

If I want to create an array of objects, I use the standard library container classes. So if I want n objects of class Option, I just put

```
std::vector<Option> v(n);
```

The memory will then be deleted automatically when necessary; in addition, copying and assignment are done for me by the std::vector class. In fact, we can think of a vector as a smart pointer owning an array of objects.

Since I never use new [], I never need delete [], and that allows the avoidance of nasty bugs caused by accidentally using the wrong version of the delete command.

In the unlikely event of needing to write a new container class, we can just use the vector class as a data member to handle the memory for us. Why is it unlikely you will need to write a container class? The standard library and Boost contain enough such classes to cover any reasonable case that is likely to arise in quant work.

Note that if we pass vector s by reference, it will be just as fast as passing pointers. The compiler inlines the data access operator, [], so we do not lose any speed on deferencing either.

Another plus of using the standard library containers over direct memory allocation is that modern versions of the standard library include range-checking in debug mode. So if you accidentally wander off the end of an array, you immediately get an exception, alerting you to a logical error instead of wondering where the silly numbers came from. Such a range-checked library ships with Visual Studio 8.0, and you can get range-checked libraries for other compilers from

http://www.stlport.org

Of course, the memory allocation commands have their uses but if you find them becoming part of your regular usage, one of the following is the case:

- you are a hard-core developer and should not be wasting your time reading a low-level book like this one;
- you have lost sight of what you should be doing, and should start focussing on numerical modelling instead.

13.6 Making the wrapper class exception safe

We have argued that we should use smart pointers since they make exception safety easy. However, we must also make sure that the smart pointers we write for ourselves are exception safe. In fact, the Wrapper class as we originally wrote it is not exception safe. Consider the assignment operator

```
Wrapper& operator=(const Wrapper<T>& original)
{
    if (this != &original)
    {
        if (DataPtr!=0)
            delete DataPtr;

        DataPtr = (original.DataPtr !=0)
                    ? original.DataPtr->clone() : 0;
    }

    return *this;
}
```

If the call to the clone method of the original object passed in throws, we have a problem. We have already deleted DataPtr so the strong guarantee is violated. But worse, any attempt to access the underlying object will be an attempt to access a dead object, and we can expect a crash. This will happen when the Wrapper goes out of scope and a second attempt is made to delete DataPtr, if not before. Thus not even the weak guarantee is satisfied.

We therefore need to recode the assignment operator in Wrapper to avoid this problem

```
Wrapper& operator=(const Wrapper<T>& original)
{
   if (this != &original)
   {
      T* newPtr = (original.DataPtr !=0) ?
                     original.DataPtr->clone() : 0;

      if (DataPtr!=0)
         delete DataPtr;

      DataPtr = newPtr;
   }

   return *this;
}
```

If the cloning throws, then the object has not been changed, so with the new design, the strong guarantee is satisfied.

13.7 Throwing in special functions

As well as throwing in ordinary code, there is the issue of what to do when an error occurs in a constructor or destructor. In this section, we look at the issues. The short version is, "it's ok to throw in a constructor but never throw in a destructor."

We examine constructors first. The main danger of throwing in a constructor is that resources acquired may not be released. The important fact to know here is that destructors are only called for fully constructed objects. So if an exception is thrown in the main body of the constructor, the destructors for all the data members are called but the destructor for the object being created is not.

So if the destructor carries out some non-trivial operations such as calling delete, we have a problem. This can be tackled in two ways. The first is simply to do any tidying up that the destructor would have done before calling throw. The second is to follow the rule of almost zero and have a trivial destructor. The second approach is much safer in that exceptions could arise in unexpected places.

One subtlety to be aware of is that the constructor of one of the data members of the class could also throw. These constructors are all called before the main routine is entered. They are called in the order that they are declared in the class declaration. On the throw, the destructors for all the objects already created will be called in reverse order. So we must also design our class data members so they will automatically delete any memory they have allocated.

What about destructors? Suppose we write a destructor for a class A that throws when it's unhappy. We let B have a data member of type A. Now consider the following snippet

```
bool flag = true;
try
{
    {
        B testObject;

        if (flag)
            throw("flag is true");
    }
}
catch(...)
{
}
```

When the throw is called, the stack is unwound and the object of type B is destroyed. As part of this is destruction, its data-member of type A is destroyed. If the destructor of A throws, the application terminates, i.e. crashes. This is specified in the C++ standard; the reason being that the compiler will not know which exception to deal with.

So never ever throw in a destructor.

13.8 Floating point exceptions

Our discussion so far has looked at C++ exceptions. There is, however, an additional source of exceptions when working with numerical code: the floating point exception. This section although important lies outside the C++ standard, and I am therefore going to restrict to discussing purely what happens with Visual Studio 8.0. For example, consider the following code

```
double x=0;
double y=1e6;
double z = y/x;
std::cout << z;
```

The default behaviour with Visual Studio is to output 1#INF. However, it would be nice to have an exception thrown at the moment such a problematic operation occurs rather than realizing at some point much later on that the computation became garbage halfway through.

It is possible to enable floating point exceptions. In this section, we discuss how to do this and how to catch them. Enabling floating point exceptions is in fact rather easy, one simply includes `Float.h` and the line of code

```
_controlfp(_EM_INEXACT,_MCW_EM);
```

Note that this is a run-time command so you can decide at run time whether you want floating point exceptions to be thrown. For example, it can be useful to switch them off when "float underflow" errors are being generated – these generally result from numbers being too small; however, too small numbers often have zero impact on the final result.

The only problem with the `__controlfp` command is that a "structured exception" is generated not a C++ exception. The effect of this is that you get an unhandled exception error even if you put a catch-all statement immediately after the offending line, and the program crashes.

To get a C++ exception, we have to call another command `_set_se_translator` defined in `Windows.h`. This tells the compiler how to translate structured expections into C++ exceptions. It takes as argument a function to be called when a structured exception is thrown. Note that the file `Windows.h` does not ship with Visual Studio Express 8.0 and you will have to install the free Microsoft Platform SDK to use it.

We illustrate its use in `FPSetup.h` and `FPSetup.cpp`.

Listing 13.1 (FPSetup.h)

```
#ifndef FP_SETUP_H
#define FP_SETUP_H

#include <Windows.h>
#include <stdexcept>

class float_exception : public std::exception {};
class fe_denormal_operand : public float_exception {};
```

```
class fe_divide_by_zero : public float_exception {};
class fe_inexact_result : public float_exception {};
class fe_invalid_operation : public float_exception {};
class fe_overflow : public float_exception {};
class fe_stack_check : public float_exception {};
class fe_underflow : public float_exception {};

void se_fe_trans_func(
    unsigned int u, EXCEPTION_POINTERS* pExp );


void EnableFloatingPointExceptions();

#endif
```

As well as declaring two functions, one that enables the exceptions and the other that declares the translation function, we declare a number of classes expressing the different sorts of exceptions that can be thrown. We do a two-level inheritance hierarchy off the standard library exception class std::exception. First we inherit the class float_exception and then we inherit all the different types of exceptions from it. Note that all these classes are simply empty classes – the information is conveyed simply by the type of object thrown rather than data contained within the object.

The upshot of this is that we can have a generic command catch-all standard library of exceptions, including floating point exceptions with catch(std:: exception), or just catch floating point exceptions with catch (float_exception), or just catch one specific type of floating point exception of choice. For example, we can use catch(fe_divide_by_zero) just to get division by zeros.

We implement these functions in FPSetup.cpp.

Listing 13.2 (FPSetup.cpp)

```
#include"FPSetup.h"
#include <Float.h>
void se_fe_trans_func( unsigned int u,
                       EXCEPTION_POINTERS* pExp )
{
    switch (u)
    {
            case STATUS_FLOAT_DENORMAL_OPERAND:
                        throw fe_denormal_operand();
```

```
                    case STATUS_FLOAT_DIVIDE_BY_ZERO:
                                   throw fe_divide_by_zero();
                    case STATUS_FLOAT_INEXACT_RESULT:
                                   throw fe_inexact_result();
                    case STATUS_FLOAT_INVALID_OPERATION:
                                   throw fe_invalid_operation();
                    case STATUS_FLOAT_OVERFLOW:
                                   throw fe_overflow();
                    case STATUS_FLOAT_UNDERFLOW:
                                   throw fe_underflow();
                    case STATUS_FLOAT_STACK_CHECK:
                                   throw fe_stack_check();

        };

    throw float_exception();
}


void EnableFloatingPointExceptions()
{
        _set_se_translator(se_fe_trans_func);
        _controlfp(_EM_INEXACT,_MCW_EM);
}
```

The implementation of se_fe_trans_func is not particularly interesting – just a switch through the different possible types of exception. One subtlety, however, is that you must change to the compiler flags to get all this to work. In particular, the /EHa flag must be set. This can be set via project properties, C/C++, code generation, enable exceptions, and should be set to "Enable C++ exceptions with SEH."

We give a simple illustration of its use in FPMain.cpp.

Listing 13.3 (FPMain.cpp)

```
#include "FPSetup.h"
#include <iostream>
#include <cmath>
int main()
{
```

```
EnableFloatingPointExceptions();

try
{
    double x;
    double y;
    std::cin >> x;
    std::cin >> y;

    double z = y/x;
    double t= exp(z);
    std::cout
        << z << " " << t << "\n";


}
catch (fe_divide_by_zero&)
{
    std::cout << "div by zero\n";
}
catch (float_exception&)
{
    std::cout << "other floating point exception\n";
}
catch(...)
{
    std::cout << "exception caught\n";
};

char c;
std::cin >> c;

return 0;
};
```

If we run this and enter 0 and 1, we get a division by zero exception. If we enter 1 and $1E6$, a float overflow occurs and we get the output "other floating point exception." If you try it without the correct flags set, the exceptions will not be caught.

Note that this code also illustrates that we can `catch` an object as a member of its base class as well as a member of its own class. The rules are that an object is caught if

- the `catch` argument matches the type of the object thrown;
- the `catch` argument type is a `public` base class of the object thrown;
- the `catch` argument is a pointer and the thrown pointer can be converted to this pointer type according to the ordinary rules of pointer conversion.

Although the exception will be thrown and we may know the type of the floating point failure, we still have the issue that we want to know where it was thrown. One way to find this out is to use the debugger. In Visual Studio, in the "debug" menu there is an "exceptions" menu that allows the user to specify that execution halt in the debugger on various sorts of exceptions. By ticking these boxes, we can cause it to stop at the precise instant the problem occurs and examine the computation. Note that we can also use the call stack to go and up and down nested function calls to see where the problem arises.

13.9 Key points

In this chapter, we have looked at various issues related to making code function well in the presence of exceptions:

- Exceptions can cause memory leaks.
- The weak or basic exception safety guarantee says that a program will be in a valid state after an exception is thrown.
- The strong exception safety guarantee says that if an exception is thrown during an operation, then the program will be left in the state it was in at the start of the operation.
- Memory leaks can be avoided by the use of smart pointers.
- The rule of almost zero advises never to write code that requires non-trivial copy constructors, assignment operators, and destructors.
- Avoid the `new []`, `delete` and `delete []` commands.
- We have to take care when writing the assignment operators of smart pointers to avoid memory leaks when `new` fails.
- Floating point errors do not by default cause C++ exceptions but they can be made to do so.

13A The new wrapper class

We have made some changes to the `Wrapper` class. In particular, we have added a new constructor and rewritten the assignment operator. Here is the revised code

Listing 13.4 (`wrapper2.h`)

```
#ifndef WRAPPER_H
#define WRAPPER_H

template< class T>
class Wrapper
{
public:

    Wrapper()
    {
        DataPtr =0;
    }

    Wrapper(const T& inner)
    {
        DataPtr = inner.clone();
    }

    Wrapper(T* DataPtr_ )
    {
        DataPtr =DataPtr_;
    }

    ~Wrapper()
    {
        if (DataPtr !=0)
            delete DataPtr;
    }

    Wrapper(const Wrapper<T>& original)
    {
        if (original.DataPtr !=0)
            DataPtr = original.DataPtr->clone();
        else
            DataPtr=0;
    }

    Wrapper& operator=(const Wrapper<T>& original)
```

```
    {

        if (this != &original)
        {
            T* newPtr = (original.DataPtr !=0) ?
                original.DataPtr->clone() : 0;

            if (DataPtr!=0)
                delete DataPtr;

            DataPtr = newPtr;
        }
        return *this;
    }

    T& operator*()
    {
        return *DataPtr;
    }

    const T& operator*() const
    {
        return *DataPtr;
    }

    const T* const operator->() const
    {
        return DataPtr;
    }

    T* operator->()
    {
        return DataPtr;
    }


private:
    T* DataPtr;
};
#endif
```

We illustrate its use in WrapperMain.cpp.

Listing 13.5 (WrapperMain.cpp)

```
//
//              requires PayOff3.cpp

#include <iostream>
#include <Wrapper2.h>
#include <PayOff3.h>

int main()
{
        double S;
        double K1,K2,K3;

        std::cout << " spot\n";
        std::cin >> S;

        std::cout << "strike1\n";
        std::cin >> K1;

        std::cout << "strike2\n";
        std::cin >> K2;

        PayOffCall one(K1);
        PayOffPut two(K2);

        PayOff* p = one.clone();
        Wrapper<PayOff> four = p;

        {
                PayOff* q = two.clone();
                Wrapper<PayOff> five = q;

                std::cout << "four :";
                std::cout << (*four)(S)
                        << " five :"
                        << (*five)(S) << "\n";

                four = five;
```

```
    }
    std::cout <<  " four :" << (*four)(S) << "\n";

    char c;
    std::cin >> c;
    return 0;
}
```

14

Templatizing the factory

14.1 Introduction

The factory pattern discussed in Chapter 10 allowed us a method of turning inputs into objects from a generalized hierarchy. It also allowed us to add extra objects without modifying any files. This is such a useful pattern that I use it all the time in all sorts of contexts. As such it is a natural candidate for templatization. The objective in this chapter is to develop such a templatized factory. This will raise additional problems regarding reusability, and we will develop new techniques to solve them.

14.2 Using inheritance to add structure

A key part of our factory was the singleton, and a key part of the singleton was the fact that it could not be copied. We achieved this by making all the constructors including the copy constructor `private`. Whilst a class will inevitably own its own constructors, there is a way of implementing a general solution to the copying problem. This will have the main virtue that a user of the class can immediately see that it cannot be copied rather than having to be told by the compiler or by inspecting to see whether a `private` copy constructor has been declared.

The key observation that makes this technique work is that if a class has a `private` copy constructor or assignment operator, then any class inherited from it cannot be copied or assigned either, since the inherited class implicitly holds a base class object.

Here's the important part of the relevant file from `Boost`, "`boost/noncopyable.hpp`".

```
class noncopyable
  {
  protected:
```

```
      noncopyable() {}
      ~noncopyable() {}
   private:   // emphasize the following members are private
      noncopyable( const noncopyable& );
      const noncopyable& operator=( const noncopyable& );
   };
```

The class is very small and has no data members. Its constructor and destructor are protected to ensure that it can only be constructed by an inherited class. The copy constructor and assignment operator are, of course, `private` to ensure that even inherited classes cannot make a copy.

To use this class we simply inherit our new class from it. For example, a typical use would be

```
class MySingleton : private noncopyable
{

...

};
```

Note that here we have used `private` inheritance for the first time. Whereas `public` inheritance express the "is a" relationship, `private` inheritance is said to express "implemented in terms of a." The main difference is that the inherited class's public interface does not contain the `public` part of the base class with `private` inheritance. In our example, of course, the base class does not contain any `public` methods, so the difference is purely in how we communicate our intentions to clients of `Singleton`.

Note that `private` inheritance and object composition are very similar. Indeed, if we had a data-member of type `noncopyable` (which is not possible unless we modify the class to have a `public` constructor), then `Singleton` would not be copyable either. However, the clear statement at the start of the class is more transparent to users.

There is also a subtlety relating to object size. If we take `sizeof` a class with no data members, what do we get? With good reason, the standard requires the number to be greater than zero. Much code implicitly assumes that this is the case. For example, suppose we created a `vector` of objects from a class with no data-members. The implementer may well have decided to multiply and divide by `sizeof` when working out an offset in memory; if zero size objects were allowed, this would cause serious trouble.

So an object of type `noncopyable` has positive size. This means that having a data member of its type will increase the size of an object; this is certainly a disincentive to using this pattern. The reader is probably now saying, "ah but the same applies to inheritance." Generally, an object contains an instance of the base class so inheriting off something simply adds to the size of the base class. However, this is not true when the base class is empty. In that case, the standard allows the compiler to optimize away the dummy space. So the private inheritance model has no overhead and the virtue of clarity.

14.3 The curiously recurring template pattern

Now suppose we decide to implement a reusable singleton via inheritance. All our singleton classes will inherit off a class called Singleton. We can inherit a Singleton class off noncopyable and make the constructor private to ensure that only methods of the class can create objects from it.

The trickier problem is how to implement the method that returns the sole instance of the class. Our previous solution was to have a method that contained a static data declaration of the class type. But the class does not know the type of the inherited class – information flows downwards not upwards with inheritance.

One solution to this problem called the "curiously recurring template pattern" is to templatize on the *type of the inherited class*. A little surprisingly, this is legal C++. Our singleton therefore takes the form

```
template<class T>
class Singleton  : private noncopyable
{
public:
    static T& Instance()
    {
        static T one;
        return one;
    }

protected:
        Singleton() {}

};
```

To create a new class which is a singleton class called MyFactory, we then code

```
class MyFactory : public Singleton<MyFactory>
{

...
```

```
private:

    ...

    MyFactory(){}
    friend class Singleton<MyFactory>;

};
```

Note the friend declaration, the constructor for `MyFactory` is `private` so we need this to allow the `Singleton` class to create the one object from `MyFactory`.

14.4 Using argument lists

Did you try Exercise 10.2? If not, think about it for a little while before continuing. In fact, let's consider a harder problem. Suppose we want to be able to specify lots of different numbers of arguments of varying types. Our first exotic option might require two arrays, a double, and three strings. Our second might require one array, a boolean, a matrix, and one string. This could happen, and if we want to implement a generic factory, then we have to cope not just with objects from the same inheritance hierarchy requiring such varying arguments, but also with pieces of code that have totally different requirements.

We could templatize on the argument type but would then still have the problem of variable numbers of arguments. We would therefore have to have a template for each possible number of arguments. This would be tiresome at best. An alternative approach is to have one class that encapsulates all reasonable sorts of arguments. Such a class is generally called an argument list. (There is a mechanism in C for functions to have variable numbers of arguments, but this is rarely if ever used in C++.)

We present such a class in `ArgList.h` from the `xlw` project. (For further discussion of `xlw` see Chapter 15.)

Listing 14.1 (`ArgList.h`)

```
#ifndef ARG_LIST_H
#define ARG_LIST_H

#include <xlw/port.h>
#include "CellMatrix.h"
#include "MyContainers.h"
#include <map>
```

```
#include <string>
#include <vector>

void MakeLowerCase(std::string& input);

class ArgumentList
{
public:

    ArgumentList(CellMatrix cells,
                    std::string ErrorIdentifier);

    ArgumentList(std::string name);


    enum ArgumentType
    {
        string, number, vector, matrix,
                boolean, list, cells
    };

    std::string GetStructureName() const;

    const std::vector<std::pair<std::string, ArgumentType> >&
                            GetArgumentNamesAndTypes()
                                const;

    std::string GetStringArgumentValue(
                                const std::string&
                                    ArgumentName);

    unsigned long GetULArgumentValue(
                                const std::string&
                                    ArgumentName);

    double GetDoubleArgumentValue(const std::string&
                                    ArgumentName);

    MyArray GetArrayArgumentValue(const std::string&
                                    ArgumentName);
```

```
MyMatrix GetMatrixArgumentValue(const std::string&
                                    ArgumentName);


bool GetBoolArgumentValue(const std::string&
                                    ArgumentName);


CellMatrix GetCellsArgumentValue(const std::string&
                                    ArgumentName);


ArgumentList GetArgumentListArgumentValue(
                            const std::string&
                                    ArgumentName);


// bool indicates whether the argument was found
bool GetIfPresent(const std::string& ArgumentName,
            unsigned long& ArgumentValue);
bool GetIfPresent(const std::string& ArgumentName,
            double& ArgumentValue);
bool GetIfPresent(const std::string& ArgumentName,
            MyArray& ArgumentValue);
bool GetIfPresent(const std::string& ArgumentName,
            MyMatrix& ArgumentValue);
bool GetIfPresent(const std::string& ArgumentName,
            bool& ArgumentValue);
bool GetIfPresent(const std::string& ArgumentName,
            CellMatrix& ArgumentValue);
bool GetIfPresent(const std::string& ArgumentName,
            ArgumentList& ArgumentValue);


bool IsArgumentPresent(const std::string&
                                ArgumentName) const;
void CheckAllUsed(const std::string& ErrorId) const;
CellMatrix AllData() const;


// data insertions

void add(const std::string& ArgumentName,
            const std::string& value);
void add(const std::string& ArgumentName,
            double value);
```

```
    void add(const std::string& ArgumentName,
             const MyArray& value);
    void add(const std::string& ArgumentName,
             const MyMatrix& value);
    void add(const std::string& ArgumentName,
             bool value);
    void add(const std::string& ArgumentName,
             const CellMatrix& values);

    void add(const std::string& ArgumentName,
             const ArgumentList& values);

    void addList(const std::string& ArgumentName,
             const CellMatrix& values);
private:

    std::string StructureName;

    std::vector<std::pair<std::string,
                ArgumentType> > ArgumentNames;
    std::map<std::string,double> DoubleArguments;
    std::map<std::string,MyArray> ArrayArguments;
    std::map<std::string,MyMatrix> MatrixArguments;
    std::map<std::string,std::string> StringArguments;
    std::map<std::string,CellMatrix> ListArguments;
    std::map<std::string,CellMatrix> CellArguments;
    std::map<std::string,bool> BoolArguments;
    std::map<std::string,ArgumentType> Names;
    std::map<std::string,bool> ArgumentsUsed;
    void GenerateThrow(std::string message,
                       unsigned long row,
                       unsigned long column);

    void UseArgumentName(const std::string& ArgumentName);
         // private as no error checking performed

    void RegisterName(const std::string& ArgumentName,
                      ArgumentType type);
};
#endif
```

The class allows data from seven different types: string, number, vector, matrix, boolean, list, and cells. An arbitrarily large amount of data from each of these types is allowed. Data are retrieved by using a `string` as key. Most of these types are self-explanatory, but we discuss the others. The type `number` is essentially a `double` but could be interpreted in others ways, e.g. an `int` or an `unsigned long`.

The type `cells` expresses the notion of a table of values that can be of multiple types: string, number, boolean, error, or empty. The idea here is to abstract the notion of values contained in cells in a spreadsheet, since ultimately xlw is a package for interfacing C++ with Excel.

The type `list` says that the argument is an `ArgList` itself. This turns out to be very useful. For example, suppose we wish to implement a random number generator factory with the arguments and type of generator specified via an `ArgList`. Now suppose we want to do a class that does anti-thetic sampling of an arbitrary generator. We make the constructor of the anti-thetic sampling decorator class (see Chapter 6) take in an `ArgList` with an argument of type `list` called "InnerGenerator" and use this to create the object to be decorated.

The idea is that the user of a factory should pass in an argument list that contains all the data necessary to identify the object to be created and to create it. These data will vary from class to class both in amount and type. The argument list has a name that expresses the base class type, e.g. "payoff," rather than the specific inherited class. The factory will retrieve the identifier key by asking the argument list for a `string` argument called "name."

The argument list is then passed by the factory to the constructor for the inherited class identified. It then queries the argument list for each piece of data it needs to create the object. Often a constructor will have to deal with optional data, so the argument list will have to provide facilities for checking if arguments are present and also for obtaining a list of the arguments' names and types. One big problem that can arise with optional arguments is that the user misspells an optional name and the constructor thinks it is not there, so we also have to introduce a mechanism for checking that all arguments have been used.

The methods of the class are in a number of types. We include two constructors, a few methods for retreiving general information, a large number of methods for retreiving information of each type, and methods for adding information of each type. We discuss these individually.

The method

```
std::string GetStructureName() const;
```

returns the type of the structure. This will always be in lower case, as will all `string`s returned by the structure.

The ability to get a complete list of all arguments by name and type is given by

```
const std::vector<std::pair<std::string, ArgumentType> >&
                        GetArgumentNamesAndTypes() const;
```

For each type we provide a method for retrieving arguments of that type. For example, to get an argument we use

```
MyArray GetArrayArgumentValue(const std::string& ArgumentName);
```

The `ArgumentName` is turned into lower case before being matched to the list of arguments. The `ArgList` class will `throw` an error if the argument requested is not present.

Note that we provide two ways of retrieving arguments of type number

```
unsigned long GetULArgumentValue(
                    const std::string& ArgumentName);
```

```
double GetDoubleArgumentValue(
                    const std::string& ArgumentName);
```

The idea here is that, whilst the data passed in from a spreadsheet will not make a distinction, it is sometimes convenient to have the data already cast to the right type before use.

To cope with optional arguments, we include

```
bool GetIfPresent(const std::string& ArgumentName,
                  MyArray& ArgumentValue);
```

Note that these methods all have the same name, since the argument type is part of the function signature. If the argument is present, `ArgumentValue` is overwritten, otherwise it is left alone. The return value indicates whether the argument was found.

We also include

```
bool IsArgumentPresent(const std::string& ArgumentName) const;
```

for testing whether an argument has been included.

To make sure no arguments have been accidentally ignored, we also include the method

```
void CheckAllUsed(const std::string& ErrorId) const;
```

which `throws` if there is an unused argument and attaches the `ErrorId` to the error message which also identifies the name of the unused argument. Clearly, we could alternatively have this method return a `bool` and then decide what to do if there is an unused argument. The reason for simply having a `throw` is to avoid the necessity of writing the same code saying to `throw` if there is an unused argument each time. Note that the reason that the data retrieval mechanisms are non-`const` is that they record whether an argument has been used.

We provide two ways of creating `ArgumentLists`

```
ArgumentList(std::string name);
```

```
ArgumentList(CellMatrix cells,
             std::string ErrorIdentifier);
```

The first of these simply creates an empty object with nothing but a name. The second is intended to be used when creating argument lists from a spreadsheet. The user will put the data in a table of cells in the sheet, and pass this table to a function. The function will then create an argument list from it; this list can then be used to create objects from a factory. We defer further discussion of this constructor to Section 14.8 where we discuss the `CellMatrix` class.

Clearly, an empty argument list is not much use so we have the ability to add arguments using the `add` methods. These are pretty self-explanatory. Note that we have an extra one `addList` which allows the addition of a `list` argument expressed by a `CellMatrix` instead of as an `ArgumentList` – we discuss the reason for this below.

An additional advantage of the add method is that when working with a function in a spreadsheet, we often want to run many cases varying one parameter. We can then use a `CellMatrix`, which describes all arguments except one, which is inputted separately and varies from cell to cell. Note that both name and value could be inputted independently allowing extra flexibility.

Our remaining `public` method is

```
CellMatrix AllData() const;
```

which converts the argument list into a `CellMatrix`, which can then be used to construct the list again if so desired. This can be useful for returning large amounts of data to a spreadsheet, or simply for checking whether the class is working correctly.

14.5 The private part of the `ArgumentList` class

How do we implement the `ArgumentList` class? We have a `std::string` that represents the structure name as a data member. We have a data member for each type which is implemented using the `std::map` class from the Standard Template Library.

This stores each data member together with its key. We store the key in lower case and will convert any keys to lower case before querying the map. Note that for all types except `list` the type stored in the `map` is the same as the type to be returned.

For `list`, we store a `CellMatrix` instead. The reason for this is that the alternative is to have type `ArgumentList` as a template argument for a `map` which is a data member of type `ArgumentList`. Whilst some compilers can cope with this, some cannot. And coping is largely dependent on the particular implementation of the standard library in use. So to avoid trouble we store `lists` as a `CellMatrix`; since we have methods to convert to and from `CellMatrix`, this is easy.

Note that there would be an inefficiency here if the `ArgumentList` was being queried repeatedly in a tight loop, but the class is not designed for efficiency in a numerical situation. Instead, it should be used for setting up objects before numerical work starts, and returning data when it is finished. Note that this is why we have included a method `addList` for adding a `CellMatrix` describing a `list`; this allows us to add a `list` without doing all these conversions.

We have two further maps. The data member

```
std::map<std::string,ArgumentType> Names;
```

stores all the names in a `map` to allow the retrieval of the type of each argument.

The data member

```
std::map<std::string,bool> ArgumentsUsed;
```

is initially set to have all `bools` `false`. Each time an argument is queried, we can then set the relevant `bool` to `true`.

We have one more data member

```
std::vector<std::pair<std::string, ArgumentType> >
                                    ArgumentNames;
```

This is simply used for storing all the names and types of arguments.

We also have three `private` methods. These are `private` since they are only to be used internally by the class.

```
void GenerateThrow(std::string message,
                   unsigned long row,
                   unsigned long column);
```

is used by the constructor that takes in a `CellMatrix`. It throws a message that identifies where the problem was in the inputted `CellMatrix`. This is to avoid code duplication between methods.

The method

```
void UseArgumentName(const std::string& ArgumentName);
```

is to set the relevant `bool` in the `ArgumentsUsed` map to `true`. Once again, we use a method here to avoid duplication.

The method

```
void RegisterName(const std::string& ArgumentName,
                  ArgumentType type);
```

updates the `ArgumentNames` and `ArgumentUsed`. This will be used by each of the add methods. It also checks that the same name has not been inserted twice.

14.6 The implementation of the `ArgumentList`

We present the source file for the `ArgumentList`, excluding the parts for conversion to and from `CellMatrix`, which we defer to Section 14.8.

Listing 14.2 (`ArgList.cpp`)

```cpp
#include "ArgList.h"
#include <algorithm>
#include <sstream>

namespace
{
    template<class T>
    T maxi(T i, T j)
    {
        return i > j ? i : j;
    }
}

void MakeLowerCase(std::string& input)
{
    std::transform(input.begin(),input.end(),input.begin(),
        tolower);
}

std::string ConvertToString(double Number)
{
    std::ostringstream os;
    os << Number;
```

```
    return os.str();

}


std::string ConvertToString(unsigned long Number)
{
    std::ostringstream os;
    os << Number;
    return os.str();

}



void ArgumentList::add(const std::string& ArgumentName,
                       const std::string& value)
{
    ArgumentType thisOne = string;
    std::pair<std::string, ArgumentType> thisPair(ArgumentName,
                        thisOne);
    ArgumentNames.push_back(thisPair);

    std::pair<std::string,std::string> valuePair(ArgumentName,
                        value);
    StringArguments.insert(valuePair);

    RegisterName(ArgumentName, thisOne);
}

void ArgumentList::add(const std::string& ArgumentName,
                       double value)
{
    ArgumentType thisOne = number;
    std::pair<std::string, ArgumentType>
                                    thisPair(ArgumentName,
                                             thisOne);
    ArgumentNames.push_back(thisPair);

    std::pair<std::string,double> valuePair(ArgumentName,
                                             value);
    DoubleArguments.insert(valuePair);
```

```
        RegisterName(ArgumentName, thisOne);


}

void ArgumentList::add(const std::string& ArgumentName,
                       const MyArray& value)
{
        ArgumentType thisOne = vector;
        std::pair<std::string,
            ArgumentType> thisPair(ArgumentName,thisOne);

        ArgumentNames.push_back(thisPair);
        ArrayArguments.insert(std::make_pair(ArgumentName,value));

        RegisterName(ArgumentName, thisOne);
}

void ArgumentList::add(const std::string& ArgumentName,
                       const MyMatrix& value)
{
        ArgumentType thisOne = matrix;
        std::pair<std::string, ArgumentType>
                                    thisPair(ArgumentName,
                                                thisOne);

        ArgumentNames.push_back(thisPair);
        MatrixArguments.insert(std::make_pair(ArgumentName,
                                            value));

        RegisterName(ArgumentName, thisOne);
}
void ArgumentList::add(const std::string& ArgumentName,
                       bool value)
{
        ArgumentType thisOne = boolean;
        std::pair<std::string, ArgumentType>
                                    thisPair(ArgumentName,
                                                thisOne);

        ArgumentNames.push_back(thisPair);
```

```
    BoolArguments.insert(std::make_pair(ArgumentName,value));

    RegisterName(ArgumentName, thisOne);
}


void ArgumentList::add(const std::string& ArgumentName,
                       const CellMatrix& values)
{
    ArgumentType thisOne = cells;
    std::pair<std::string, ArgumentType>
                                    thisPair(ArgumentName,
                                             thisOne);

    ArgumentNames.push_back(thisPair);
    CellArguments.insert(std::make_pair(ArgumentName,
                                        values));

    RegisterName(ArgumentName, thisOne);

}


void ArgumentList::addList(const std::string& ArgumentName,
                       const CellMatrix& values)
{
    ArgumentType thisOne = list;
    std::pair<std::string, ArgumentType>
                        thisPair(ArgumentName,thisOne);
    ArgumentNames.push_back(thisPair);
    ListArguments.insert(std::make_pair(ArgumentName,values));

    RegisterName(ArgumentName, thisOne);
}

void ArgumentList::add(const std::string& ArgumentName,
                       const ArgumentList& values)
{
    CellMatrix cellValues(values.AllData());
    addList(ArgumentName,cellValues);
}
```

```cpp
void ArgumentList::RegisterName(const std::string&
                                ArgumentName,ArgumentType type)
{
    ArgumentNames.push_back(std::make_pair(ArgumentName,type));

    if (!(Names.insert(*ArgumentNames.rbegin()).second) )
        throw("Same argument name used twice"+ArgumentName);

    ArgumentsUsed.insert(std::pair<std::string,bool>
                                (ArgumentName,false));
}

std::string ArgumentList::GetStructureName() const
{
    return StructureName;
}

const std::vector<std::pair<std::string,
        ArgumentList::ArgumentType> >&
                ArgumentList::GetArgumentNamesAndTypes() const
{
    return ArgumentNames;
}

void ArgumentList::UseArgumentName(const std::string&
                                            ArgumentName)
{
    std::map<std::string,bool>::iterator it=
                                ArgumentsUsed.find(ArgumentName);
    it->second =true;
}

std::string ArgumentList::GetStringArgumentValue(const
                                std::string& ArgumentName_)
{
    std::string ArgumentName(ArgumentName_);
    MakeLowerCase(ArgumentName);
    std::map<std::string, std::string>::const_iterator
                it = StringArguments.find(ArgumentName);
```

```
    if (it == StringArguments.end())
        throw(StructureName+std::string(" unknown string"
                "argument asked for :")+ArgumentName);


    UseArgumentName(ArgumentName);


    return it->second;


}


unsigned long ArgumentList::GetULArgumentValue(const
                            std::string& ArgumentName_)
{
    std::string ArgumentName(ArgumentName_);
    MakeLowerCase(ArgumentName);
    std::map<std::string, double>::const_iterator
                it = DoubleArguments.find(ArgumentName);


    if (it == DoubleArguments.end())
        throw(StructureName+std::string(
            " unknown unsigned long argument asked for :")
                +ArgumentName);


    UseArgumentName(ArgumentName);


    return static_cast<unsigned long>(it->second);
}


double ArgumentList::GetDoubleArgumentValue(const
                            std::string& ArgumentName_)
{
    std::string ArgumentName(ArgumentName_);
    MakeLowerCase(ArgumentName);
    std::map<std::string, double>::const_iterator
                it = DoubleArguments.find(ArgumentName);


    if (it == DoubleArguments.end())
        throw(StructureName+std::string(
                    " unknown double argument asked for :")
                    +ArgumentName);
```

```
    UseArgumentName(ArgumentName);
    return it->second;
}


MyArray ArgumentList::GetArrayArgumentValue(const
                                std::string& ArgumentName_)


{
    std::string ArgumentName(ArgumentName_);
    MakeLowerCase(ArgumentName);
    std::map<std::string, MyArray>::const_iterator
                    it = ArrayArguments.find(ArgumentName);

    if (it == ArrayArguments.end())
        throw(StructureName+std::string(
                        " unknown array argument asked for :")
                        +ArgumentName);



    UseArgumentName(ArgumentName);
        return it->second;

}


MJMatrix ArgumentList::GetMatrixArgumentValue(const
                                std::string& ArgumentName_)
{

    std::string ArgumentName(ArgumentName_);
    MakeLowerCase(ArgumentName);
    std::map<std::string, MJMatrix>::const_iterator
                    it = MatrixArguments.find(ArgumentName);

    if (it == MatrixArguments.end())
        throw(StructureName+std::string(
                        " unknown matrix argument asked for :")
                        +ArgumentName);

    UseArgumentName(ArgumentName);
    return it->second;
```

```
}
bool ArgumentList::GetBoolArgumentValue(const
                                std::string& ArgumentName_)
{
    std::string ArgumentName(ArgumentName_);
    MakeLowerCase(ArgumentName);
    std::map<std::string, bool>::const_iterator it =
                    BoolArguments.find(ArgumentName);

    if (it == BoolArguments.end())
        throw(StructureName+std::string(
          " unknown bool argument asked for :")+ArgumentName);

    UseArgumentName(ArgumentName);
    return it->second;
}


ArgumentList ArgumentList::GetArgumentListArgumentValue(
                          const std::string& ArgumentName_)
{
    std::string ArgumentName(ArgumentName_);
    MakeLowerCase(ArgumentName);
    std::map<std::string, CellMatrix>::const_iterator
                it = ListArguments.find(ArgumentName);

 if (it == ListArguments.end())
        throw(StructureName+std::string(
                    " unknown ArgList argument asked for :")
                    +ArgumentName);
    UseArgumentName(ArgumentName);
    return ArgumentList(it->second,ArgumentName);
}


CellMatrix ArgumentList::GetCellsArgumentValue(
                        const std::string& ArgumentName_)
{
    std::string ArgumentName(ArgumentName_);
    MakeLowerCase(ArgumentName);
    std::map<std::string, CellMatrix>::const_iterator
                        it = CellArguments.find(ArgumentName);
```

```
    if (it == CellArguments.end())
        throw(StructureName+std::string(
          " unknown Cells argument asked for :")+ArgumentName);

    UseArgumentName(ArgumentName);
    return it->second;
}

bool ArgumentList::IsArgumentPresent(
                    const std::string& ArgumentName_) const
{
    std::string ArgumentName(ArgumentName_);
    MakeLowerCase(ArgumentName);
    return (Names.find(ArgumentName) != Names.end());
}

void ArgumentList::CheckAllUsed(
                        const std::string& ErrorId) const
{
    std::string unusedList;

    for (std::map<std::string,bool>::const_iterator it
     = ArgumentsUsed.begin(); it != ArgumentsUsed.end(); it++)
    {
        if (!it->second)
                    unusedList+=it->first + std::string(", ");
    }

    if (unusedList !="")
        throw("Unused arguments in "+ErrorId+" "+StructureName
                                        +" "+unusedList);

}

void ArgumentList::GenerateThrow(std::string message,
                                unsigned long row,
                                unsigned long column)
{
    throw(StructureName
                +" "+message
```

```
                +" row:"
                +ConvertToString(row)
                +"; column:"+ConvertToString(column)+".");
}


ArgumentList::ArgumentList(std::string name)
     : StructureName(name)
{

}


bool ArgumentList::GetIfPresent(
                    const std::string& ArgumentName,
                              unsigned long& ArgumentValue)
{
    if (!IsArgumentPresent(ArgumentName))
        return false;

    ArgumentValue = GetULArgumentValue(ArgumentName);
    return true;
}


bool ArgumentList::GetIfPresent(
                    const std::string& ArgumentName,
                              double& ArgumentValue)
{
    if (!IsArgumentPresent(ArgumentName))
        return false;
    ArgumentValue = GetDoubleArgumentValue(ArgumentName);
    return true;
}


bool ArgumentList::GetIfPresent(
                    const std::string& ArgumentName,
                              MyArray& ArgumentValue)
{
    if (!IsArgumentPresent(ArgumentName))
        return false;

    ArgumentValue = GetArrayArgumentValue(ArgumentName);
```

```
        return true;
}

bool ArgumentList::GetIfPresent(
                    const std::string& ArgumentName,
                                MyMatrix& ArgumentValue)
{
    if (!IsArgumentPresent(ArgumentName))
       return false;

    ArgumentValue = GetMatrixArgumentValue(ArgumentName);
    return true;
}

bool ArgumentList::GetIfPresent(
                    const std::string& ArgumentName,
                                bool& ArgumentValue)
{
    if (!IsArgumentPresent(ArgumentName))
       return false;

    ArgumentValue = GetBoolArgumentValue(ArgumentName);
    return true;
}

bool ArgumentList::GetIfPresent(
                    const std::string& ArgumentName,
                                CellMatrix& ArgumentValue)
{
    if (!IsArgumentPresent(ArgumentName))
       return false;

    ArgumentValue = GetCellsArgumentValue(ArgumentName);
    return true;
}

bool ArgumentList::GetIfPresent(
                    const std::string& ArgumentName,
                                ArgumentList& ArgumentValue)
{
```

```
  if (!IsArgumentPresent(ArgumentName))
    return false;

  ArgumentValue = GetArgumentListArgumentValue(ArgumentName);
  return true;
}
```

We start with a simple implementation of the maximum function: `maxi`. Note that the correct thing to do here is actually use the `std::max` function from the standard template library; however, some implementations "forgot" to include it (notably Visual Studio 6.0), so in the interests of cross-platform compatibility, we use an alternative.

We have three functions for manipulating strings.

```
void MakeLowerCase(std::string& input);
```

simply takes a `string` and using the `tolower` function from the C Standard Library, converts its elements to lower case. Note the use of the `transform` algorithm from the standard template library which is neater than looping through the elements of the `string`.

The two `ConvertToString` functions take in numbers and spit out `strings`. This is acheived by using the `sstream` class. This works similarly to `iostreams`. The difference being that the objective is to create a `string` rather than an in-out buffer. We include these functions since they will be useful when creating error messages that say that a certain element of a `CellMatrix` is incorrect, which is very useful when debugging spreadsheets.

We have an `add` method for each argument type. Almost all of these do the same things: add the argument name and type to `ArgumentNames`, insert the name and value into the `map` for this argument type, and call `RegisterName`. The method `addList` takes the same form.

The one `add` method that is different is the one for adding `lists`. This converts the input `ArgumentList` into a `CellMatrix` and calls the `addList` method, thus avoiding the issue of having `ArgumentList` data members.

The "get" methods are also very similar to each other. For each one, we copy the input string, convert it to lower case, and look up the `map` to see if it is present. If it is not present, we `throw`, and if it is, we store the fact that it has been used, and return the value.

Once again it is the `list` method that has an additional step. Here we get a `CellMatrix` from the `map` and convert it to an `ArgumentList` on final return. A subtlety worth mentioning is that with the current design, it is only at this point that the `CellMatrix` is checked for validity. So if the `CellMatrix` contains errors, then a `throw` will occur. Note that we could create a dummy

variable of type `ArgumentList` to be discarded in the `addList` method from the
`CellMatrix` in order to check the argument's validity at the time of addition to the
object.

We also include the `GetIfPresent` methods to make it easy for the user to
deal with optional arguments. These simply test if the argument is present and
if it is, overwrite the parameter passed by value. A `bool` is returned indicating
if the argument was found. These were included to save the user from having to
repeatedly write code to test if an argument were present and then do one thing it
it was and another if it was not.

The remaining methods are self-explanatory and we do not comment on the
implementation further.

14.7 Cell matrices

Suppose we are interfacing a function with EXCEL or another spreadsheet. The
most general form of input will be a table of values from the sheet. The object of the `CellMatrix` is to abstractize this concept. We can use this class as
a facade between the spreadsheet's internal data types and our numerical code's
objects.

The class is presented in `<xlw/CellMatrix.h>`

Listing 14.3 (`CellMatrix.h`)

```
#ifndef CELL_MATRIX_H
#define CELL_MATRIX_H

#include <xlw/port.h>
#include <string>
#include <vector>
#include <xlw/MyContainers.h>

class CellValue
{

public:
    bool IsAString() const;
    bool IsANumber() const;
    bool IsBoolean() const;
    bool IsError() const;
    bool IsEmpty() const;
```

```cpp
    CellValue(const std::string&);
    CellValue(double Number);
    CellValue(unsigned long Code, bool Error);
                //Error = true if you want an error code

    CellValue(bool TrueFalse);
    CellValue(const char* values);
    CellValue(int i);


    CellValue();

    const std::string& StringValue() const;
    double NumericValue() const;
    bool BooleanValue() const;
    unsigned long ErrorValue() const;

    std::string StringValueLowerCase() const;

    enum ValueType
    {
        string, number, boolean, error, empty
    };

    operator std::string() const;
    operator bool() const;
    operator double() const;
    operator unsigned long() const;

    void clear();
private:
    ValueType Type;
    std::string ValueAsString;
    double ValueAsNumeric;
    bool ValueAsBool;
    unsigned long ValueAsErrorCode;
};

class CellMatrix
{
```

```
public:

    CellMatrix(unsigned long rows, unsigned long columns);
    CellMatrix();
    CellMatrix(double x);
    CellMatrix(std::string x);
    CellMatrix(const char* x);
    CellMatrix(const MyArray& data);
    CellMatrix(const MyMatrix& data);
    CellMatrix(unsigned long i);
    CellMatrix(int i);

    const CellValue& operator()(
                    unsigned long i, unsigned long j) const;
    CellValue& operator()(unsigned long i, unsigned long j);

    unsigned long RowsInStructure() const;
    unsigned long ColumnsInStructure() const;

    void PushBottom(const CellMatrix& newRows);

private:

    std::vector<std::vector<CellValue> > Cells;
    unsigned long Rows;
    unsigned long Columns;
};
CellMatrix MergeCellMatrices(const CellMatrix& Top,
                            const CellMatrix& Bottom);
#endif
```

The implementation of the class is largely as a table of objects of type `CellValue`. We therefore discuss the `CellValue` class first. This is intended to represent the possible values a cell can hold. Thus we have 5 types of values:

```
enum ValueType
{
        string, number, boolean, error, empty
};
```

giving the possibilities of it holding a string, a double, a bool, an error code, or simply being empty. The error codes are represented by unsigned longs, in accordance with the practice in EXCEL.

The methods

```
bool IsAString() const;
bool IsANumber() const;
bool IsBoolean() const;
bool IsError() const;
bool IsEmpty() const;
```

allow us to test if a CellValue is of a given type. Once we know it is of that type, then we can get it via the methods

```
const std::string& StringValue() const;
double NumericValue() const;
bool BooleanValue() const;
unsigned long ErrorValue() const;
std::string StringValueLowerCase() const;
```

with the obvious effects. Note that a CellValue can be of at most one type and attempting to use it as another will yield a throw.

Note that we also have the methods

```
operator std::string() const;
operator bool() const;
operator double() const;
operator unsigned long() const;
```

which may appear a little confusing to the reader. These are implicit conversion operators. For example, suppose a routine, f, expects a double and we have a CellValue called x holding a double. We can use our original methods to code

```
f(x.NumericValue());
```

but it would be nice if we could just put

```
f(x);
```

Implicit conversion operators allow us to do this. The declaration

```
operator double() const;
```

says that the CellValue can be treated as a double and, when this is done, the method

```
CellValue::operator double() const
```

is called to give the requisite value. Note that with conversion to user-defined classes an alternative is simply to write a new constructor that takes in a `CellValue`, but this is not an option for inbuilt types such as `doubles`.

We also provide constructors for various types to make it easy to create `CellValues`. In addition, we provide a method for clearing the value: `clear()`.

The implementation of this class is straightforward and the details can be found in the file `CellMatrix.cpp` in the `xlw` project.

The `CellMatrix` class itself is just a table of values implemented as a `vector` of `vectors` for simplicity. Note that using a `vector` of `vectors` is not recommended for numerical code for efficiency reasons – it may result in data in the same matrix being in rather different parts of the memory and switching memory locations is time consuming. Here, however, we are purely interested in convenience and the design is adequate.

The main thing to remark on in the class is the number of constructors. This is because we will want to write functions returning lots of different types to the spreadsheet. By making the `CellMatrix` constructors take in all of these, we can simply write routines that return them and convert to a `CellMatrix` automatically and via that to a spreadsheet data-type. Otherwise the user will be perpetually writing code at the end of routines to convert data to the correct type.

Note we include a couple of routines for merging `CellMatrix` objects. `PushBottom` simply adds some new rows to the bottom of the `CellMatrix`, widening the object if necessary. `MergeCellMatrices` does essentially the same thing but with a non-member function interface.

The implementation of this class is straightforward and can be found in `CellMatrix.cpp`.

14.8 Cells and the `ArgumentLists`

We now return to our discussion of the `ArgumentList`, and in particular look at its methods relating to cells. The most important of these is the constructor that takes in a `CellMatrix`

```
ArgumentList(CellMatrix cells,
             std::string ErrorIdentifier);
```

The idea is that the user in a spreadsheet enters a table of values and these are then used to construct an argument list, which can then be used to create on object from the factory. The constructor takes in an additional `string` to make it easy to identify where the problem occurred in the event of an error being thrown.

The constructor uses an auxiliary routine ExtractCells. The implementation is fairly straightforward and we only comment on the unusual parts.

```
CellMatrix ExtractCells(CellMatrix& cells,
                        unsigned long row,
                        unsigned long column,
                        std::string ErrorId,
                        std::string thisName,
                        bool nonNumeric)
{
        if (!cells(row,column).IsANumber())
                throw(ErrorId+" "+thisName+
                                "rows and columns expected.");
        if (cells.ColumnsInStructure() <= column+1)
                throw(ErrorId+" "+thisName+
                                "rows and columns expected.");
        if (!cells(row,column+1).IsANumber())
                throw(ErrorId+" "+thisName+
                                "rows and columns expected.");

        unsigned long numberRows = cells(row,column);
        unsigned long numberColumns = cells(row,column+1);

        cells(row,column).clear();
        cells(row,column+1).clear();

        CellMatrix result(numberRows,numberColumns);

        if (numberRows +row+1>cells.RowsInStructure())
                throw(ErrorId+" "+thisName+
                        "insufficient rows in structure");

        if (numberColumns +column>cells.ColumnsInStructure())
                throw(ErrorId+" "+thisName+
                        "insufficient columns in structure");

        for (unsigned long i=0; i < numberRows; i++)
                for (unsigned long j=0; j < numberColumns; j++)
                {
                        result(i,j) = cells(row+1+i,column+j);
```

```
                                cells(row+1+i,column+j).clear();

                        if (!result(i,j).IsANumber())
                                nonNumeric = true;
                }


        return result;



}


ArgumentList::ArgumentList(CellMatrix cells,
                           std::string ErrorId)
{
    CellValue empty;
    unsigned long rows = cells.RowsInStructure();
    unsigned long columns = cells.ColumnsInStructure();

    if (rows == 0)
       throw(std::string(
                "Argument List requires non empty cell matix")
                      +ErrorId);

    if (!cells(0,0).IsAString())
       throw(
          std::string("a structure name must be specified"
                        "for argument list class ")+ErrorId);
    else
    {
       StructureName = cells(0,0).StringValueLowerCase();
       cells(0,0) = empty;
    }
    {for (unsigned long i=1; i < columns; i++)
        if (!cells(0,i).IsEmpty() )
           throw("An argument list should only"
                "have the structure name"
                "on the first line: "
                +StructureName+ " " + ErrorId);
```

```
}

ErrorId +=" "+StructureName;
{for (unsigned long i=1; i < rows; i++)
    for (unsigned long j=0; j < columns; j++)
        if (cells(i,j).IsError())
            GenerateThrow("Error Cell passed in ",i,j);}


unsigned long row=1UL;

while (row < rows)
{
    unsigned long rowsDown=1;
    unsigned column = 0;

    while (column < columns)
    {
        if (cells(row,column).IsEmpty())
        {
        // check nothing else in row
            while (column< columns)
            {
                if (!cells(row,column).IsEmpty())
                    GenerateThrow("data or value where"
                                    " unexpected."
                                       ,row, column);
                ++column;
            }
        }
        else // we have data
        {
            if (!cells(row,column).IsAString())
                GenerateThrow(
                    "data  where name expected.",
                        row, column);

            std::string thisName(
                    cells(row,
                  column).StringValueLowerCase());
```

```
                if (thisName =="")
                    GenerateThrow(
                            "empty name not permissible.",
                                row, column);

                if (rows == row+1)
                    GenerateThrow("No space where data"
                                "expected below name",
                                row, column);

                cells(row,column).clear();
// weird syntax to satisfy VC6
                CellValue* belowPtr = &cells(row+1,column);
                CellValue& cellBelow = *belowPtr;

                if (cellBelow.IsEmpty())
                    GenerateThrow(
                        "Data expected below name",
                                row, column);

                if (cellBelow.IsANumber())
                {
                    add(thisName, cellBelow.NumericValue());
                    column++;
                    cellBelow=empty;
                }
                else
                    if (cellBelow.IsBoolean())
                    {
                        add(thisName,
                                cellBelow.BooleanValue());
                        column++;
                        cellBelow=empty;
                    }
                    else // ok its a string
                    {
                        std::string stringVal
                                = cellBelow.
                                    StringValueLowerCase();
```

```
if ( (cellBelow.StringValueLowerCase()
        == "list") ||
    (cellBelow.StringValueLowerCase()
        == "matrix") ||
    (cellBelow.StringValueLowerCase()
        == "cells") )
{
  bool nonNumeric = false;
  CellMatrix extracted(
          ExtractCells(cells,
                       row+2,
                       column,
                       ErrorId,
                       thisName,
                       nonNumeric));


  if (cellBelow.StringValueLowerCase()
                          == "list")
  {
      ArgumentList value(extracted,
                      ErrorId+":"
                        +thisName);

      addList(thisName,
            extracted);
              //note not value

  }

  if (cellBelow.StringValueLowerCase()
                          == "cells")
  {
      add(thisName,extracted);
  }


  if (cellBelow.StringValueLowerCase()
                          == "matrix")
  {
```

```
            if (nonNumeric)
                throw("Non numerical value"
                        " in matrix argument :"
                        +thisName+ " "+ErrorId);

            MJMatrix value(
              extracted.RowsInStructure(),
              extracted.ColumnsInStructure());

            for (unsigned long i=0;
                   i < extracted.
                       RowsInStructure(); i++)
                for (unsigned long j=0;
                    j < extracted.
                      ColumnsInStructure(); j++)
                    ChangingElement(value,i,j) =
                              extracted(i,j);
                add(thisName,value);

        }

        cellBelow = empty;
        rowsDown = maxi(rowsDown,
            extracted.RowsInStructure()+2);
        column+= extracted.
            ColumnsInStructure();
    }
    else // ok its an array or boring string
    {
        if (cellBelow.StringValueLowerCase()
               == "array"
              || cellBelow.StringValueLowerCase()
               == "vector" )
        {
            cellBelow.clear();

            if (row+2>= rows)
                throw(ErrorId
                       +" data expected below"
                       "array "+thisName);
```

```
            unsigned long size =
                cells(row+2,column);
            cells(row+2,column).clear();

            if (row+2+size>=rows)
                throw(ErrorId
                        +" more data expected"
                          "below array"+thisName);



            MyArray theArray(size);

            for (unsigned long i=0; i < size;
                    i++)
            {
                theArray[i] =
                    cells(row+3+i,column);
                cells(row+3+i,column).clear();
            }

            add(thisName,theArray);

            rowsDown = maxi(rowsDown,size+2);

            column+=1;
        }
        else
        {
            std::string value =
                cellBelow.StringValueLowerCase();
            add(thisName,value);
            column++;
            cellBelow=empty;
        }
    }
  }
 }
}
row+=rowsDown+1;
```

```
       }

       {for (unsigned long i=0; i < rows; i++)
           for (unsigned long j=0; j < columns; j++)
               if (!cells(i,j).IsEmpty())
               {
                   GenerateThrow("extraneous data "+ErrorId,i,j);
               }}
}
```

The constructor takes the name for the structure the `string` in the top left corner of the `CellMatrix`. In the event, that this is not a `string`, it throws. It also throws if the rest of the line is non-empty.

Throughout, every time a cell is used, its value is set to empty. This means that, at the end, we can check that every value has been used simply by checking that all cells are empty; if one is not, we throw. We also check to ensure that no error values have been passed in, and return an error message if they have.

We then scan through each row looking for identifier tags. If we find a string, then we look for data below it. Either there is data immediately below, or there is a tag specifying type of data: "cells, list, matrix, array, vector." The types "vector" and "array" both specify an "array." If there is a tag, we look below again for the dimension of the data, and then extract out the table of data using the `ExtractCells` function.

Note that if the argument is of type "list", we convert it to an `ArgumentList` and then discard the result. This allows us to be sure that no errors will be generated by the conversion to an `ArgumentList` at a later stage when the object is queried for this argument.

If there is no tag, then the argument is a number, boolean or string and we identify the type from the `CellValue` and add it to the `ArgumentList`.

Whenever an error is found, we call the `GenerateThrow` method which attaches the row and column of the problem to the error message to make it easier for the user to spot the problem.

14.9 The template factory

We have seen how to code an argument list class and how to create objects of the class from a spreadsheet input. This solves the factory problem of having to cope with many types of arguments, since we just encapsulate them all in the new class. We are now in a position to develop the template factory advertised at the start of the chapter.

Here is the factory from `xlw`

Listing 14.4 (ArgListFactory.h)

```cpp
#ifndef ARG_LIST_FACTORY_H
#define ARG_LIST_FACTORY_H
#ifdef _MSC_VER
#if _MSC_VER < 1250
#pragma warning(disable:4786)
#define VC6
#endif
#endif

#include <xlw/ArgList.h>
#include <map>
#include <string>

template<class T>
class ArgListFactory;

// friend rather than method to avoid bug in VC6.0
// with static data in member template functions
template<class T>
ArgListFactory<T>& FactoryInstance()
{
    static ArgListFactory<T> object;
    return object;
}

template<typename T>
class ArgListFactory
{
public:
#ifndef VC6
    friend ArgListFactory<T>& FactoryInstance<>();
#else
    friend ArgListFactory<T>& FactoryInstance();
#endif
    typedef T* (*CreateTFunction)(const ArgumentList& );
    void RegisterClass(std::string ClassId, CreateTFunction);
    T* CreateT(ArgumentList args);
    ~ArgListFactory(){};
```

```
private:
   std::map<std::string, CreateTFunction> TheCreatorFunctions;
   std::string KnownTypes;
   ArgListFactory(){}
   ArgListFactory(const ArgListFactory&){}
   ArgListFactory& operator=(
                    const ArgListFactory&){ return *this;}
};

template<typename T>
void ArgListFactory<T>::RegisterClass(std::string ClassId,
                               CreateTFunction CreatorFunction)
{
   MakeLowerCase(ClassId);
   TheCreatorFunctions.insert(
          std::pair<std::string,
             CreateTFunction>(ClassId,CreatorFunction));
   KnownTypes+=" "+ClassId;
}

template<typename T>
T* ArgListFactory<T>::CreateT(ArgumentList args)
{

   std::string Id = args.GetStringArgumentValue("name");

   if  (TheCreatorFunctions.find(Id) ==
          TheCreatorFunctions.end())
   {
      throw(Id+" is an unknown class. Known types are"
            +KnownTypes);
   }

   return (TheCreatorFunctions.find(Id)->second)(args);
}


// easy access function
template<class T>
T* GetFromFactory(const ArgumentList& args)
```

```
{
    return FactoryInstance<T>().CreateT(args);
}
#endif
```

The factory is templatized on a type T, which, is the type of the base class. After studying how to do generic singletons at the start of the chapter, you will note that this is not how the factory has been done. The reason is that the curiously recurring template pattern singleton is too smart for some compilers (e.g. VC6.0); when optimizing they get confused about how many copies there are of a `static` variable declared in a method of a template class.

Instead, we therefore work with a `friend` function called `FactoryInstance` which can access the `private` constructor. This has a `static` variable of type `ArgListFactory<T>`, and so plays the same role as the `static` member function `Instance` in our previous factory. Note the syntax for the `friend` declaration:

```
#ifndef VC6
    friend ArgListFactory<T>& FactoryInstance<>();
#else
    friend ArgListFactory<T>& FactoryInstance();
#endif
```

In up-to-date compilers, we have to include <> at the end of the method name.

The rest of the template class is very similar to our non-template factory. Note that we make the names lower case to avoid confusion, and we store a list of all names registered to make it easy to guide the user when an invalid name is passed in. The `CreateT` method only takes in an `ArgumentList` and does not take in a separate key; instead the key is queried from the `ArgumentList` with the tag "name."

We include the extra function

```
T* GetFromFactory(const ArgumentList& args)
```

to make calling the factory particularly easy.

We also need the helper class to register classes inherited from T with the factory. This is done in the file `xlw/ArgListFactoryHelper.h`

```
#ifndef ARG_LIST_FACTORY_HELPER_H
#define ARG_LIST_FACTORY_HELPER_H
#include <xlw/ArgListFactory.h>
#include <string>

template<class TBase, class TDerived>
```

```
class FactoryHelper
{
public:
  FactoryHelper(std::string);
  static TBase* create(const ArgumentList&);
  ~FactoryHelper(){}
};


template<class TBase, class TDerived>
FactoryHelper<TBase,TDerived>::FactoryHelper(std::string id)
{
    MakeLowerCase(id);
    FactoryInstance<TBase >().RegisterClass(id,
                                  FactoryHelper<TBase,
                                  TDerived>::create);
}


template<class TBase, class TDerived>
TBase*
FactoryHelper<TBase,TDerived>::create(
                                  const ArgumentList& Input)
{
    return new TDerived(Input);
}

#endif
```

Here everything has been templatized on both the base class and the derived class.
The derived class will be the class being registered and the base class will specify
the factory to be registered with. Otherwise, our helper class is very much the same
as the non-template one.

Note that we have written our class to only work with the ArgumentList; we
could go further and templatize on the argument type, then having two template parameters for the factory and three for the helper class. However, the ArgumentList
class is sufficiently general that the extra flexibility would seem to gain us little at
the cost of opaque syntax.

14.10 Using the templatized factory

We have now achieved our objective; we have a general template factory, which will take in multiple arguments. We return to our original motivating example: coding a factory for a pay-off class that takes in multiple arguments. An example of this is given in the "TestFiles" folder in `xlw`.

The `PayOff` class there is very simple

Listing 14.5 (`PayOff.h`)

```
#ifndef PAYOFF_H
#define PAYOFF_H
class PayOff
{
public:

    PayOff();
    virtual double operator()(double Spot) const=0;
    virtual ~PayOff();
    virtual PayOff* clone() const=0;

private:

};
#endif
```

The trivial implementations of the constructor and destructor are in `PayOff.cpp`. Inherited from this class, we have three examples given in `PayOffConcrete.cpp`.

Listing 14.6 (`PayOffConcrete.h`)

```
#ifndef PAYOFF_CONCRETE_H
#define PAYOFF_CONCRETE_H

#include "PayOff.h"
#include <xlw/ArgList.h>
#include <xlw/Wrapper.h>
#include <xlw/ArgListFactory.h>

class PayOffCall : public PayOff
{
public:
```

```cpp
    PayOffCall(ArgumentList args);
    virtual double operator()(double Spot) const;
    virtual ~PayOffCall(){}
    virtual PayOff* clone() const;

private:
    double Strike;

};


class PayOffPut : public PayOff
{
public:

    PayOffPut(ArgumentList args);
    virtual double operator()(double Spot) const;
    virtual ~PayOffPut(){}
    virtual PayOff* clone() const;

private:
    double Strike;

};


class PayOffSpread : public PayOff
{
public:

    PayOffSpread(ArgumentList args);
    virtual double operator()(double Spot) const;
    virtual ~PayOffSpread(){}
    virtual PayOff* clone() const;

private:
    Wrapper<PayOff> OptionOne;
    Wrapper<PayOff> OptionTwo;
    double Volume1;
    double Volume2;
```

```
};
```

```
#endif
```

We have classes for the put, the call, and a spread, which is a linear multiple of two other pay-offs. In all three cases, the sole constructor takes an `ArgumentList`, so the factory is directly usable. The class `PayOffSpread` can be viewed as an example of the *composite* pattern, which is similar to decorator; the difference being that more than one underlying class is involved.

The classes are implemented in `PayOffConcrete.cpp`.

Listing 14.7 (`PayOffConcrete.cpp`)

```cpp
#include <xlw/port.h>
#include "PayOffConcrete.h"

PayOffCall::PayOffCall(ArgumentList args)
{
    if (args.GetStructureName() != "payoff")
      // must be lower case here throw("payoff structure expected
        in PayOffCall class");

    if (args.GetStringArgumentValue("name") != "call")
        throw("payoff list not for call passed to PayOffCall"
              " : got "+args.GetStringArgumentValue("name"));

    Strike = args.GetDoubleArgumentValue("strike");
    args.CheckAllUsed("PayOffCall");
}

double PayOffCall::operator () (double Spot) const
{
    return Spot-Strike > 0.0 ? Spot-Strike   :0.0;
}

PayOff* PayOffCall::clone() const
{
    return new PayOffCall(*this);
}
double PayOffPut::operator () (double Spot) const
```

```
{
    return Strike-Spot > 0.0 ? Strike-Spot     :0.0;
}


PayOffPut::PayOffPut(ArgumentList args)
{
if (args.GetStructureName() != "payoff")
    // must be lower case here throw("payoff structure expected"
        "in PayOffCall class");

    if (args.GetStringArgumentValue("name") != "put")
        throw("payoff list not for put passed to PayOffPut : got "
                +args.GetStringArgumentValue("name"));

    Strike = args.GetDoubleArgumentValue("strike");
    args.CheckAllUsed("PayOffPut");
}


PayOff* PayOffPut::clone() const
{
    return new PayOffPut(*this);
}



double PayOffSpread::operator()(double Spot) const
{
    return Volume1*(*OptionOne)(Spot)+
                                Volume2*(*OptionTwo)(Spot);
}
PayOffSpread::PayOffSpread(ArgumentList args)
{
    if (args.GetStructureName() != "payoff")
    // must be lower case here throw("payoff structure expected"
        "in PayOffCall class");

    if (args.GetStringArgumentValue("name") != "spread")
        throw("payoff list not for spread passed to"
                "payoffspread : got"+args.GetStringArgumentValue(
                                                "name"));
```

```
if (!args.GetIfPresent("Volume1",Volume1))
    Volume1= 1.0;

if (!args.GetIfPresent("Volume2",Volume2))
    Volume2= -1.0;

OptionOne = Wrapper<PayOff>(GetFromFactory<PayOff>(
                args.GetArgumentListArgumentValue(
                                "optionone")));

OptionTwo = Wrapper<PayOff>(GetFromFactory<PayOff>(
                args.GetArgumentListArgumentValue(
                                "optiontwo")));

    args.CheckAllUsed("PayOffSpread");
}


PayOff* PayOffSpread::clone() const
{
    return new PayOffSpread(*this);
}
```

The implementation of these classes are straightforward, with the only interest being in how the `ArgumentList` class is used. For the `PayOffPut` class, we first check that the `ArgumentList` class has been tagged with "payoff." Note that all data passed in have been put into lower case so we must use lower case when checking. We then check that the name argument is indeed "put." In each case, we throw if there is a problem.

We get the strike by calling `GetDoubleArgumentValue("strike")` and put it into the relevant data member. Finally, we make sure that the user has not supplied extra irrelevant arguments using the `CheckAllUsed` method.

The constructor for `PayOffSpread` is more interesting. The first part is the same as before. We then check to see if the notionals of the two underlying options have been specified and otherwise set default values.

To get the underlying options themselves, we use list arguments from the `ArgumentList` passed in and call the same factory. Note that the factory returns raw pointers, but these are immediately taken over by the `Wrapper` class, which ensures that they are properly memory managed.

Note the important synergies here between the composite pattern and the `ArgumentList` class. We are able to bring our composite into the factory because it is legitimate to have data stored in the `ArgumentList` class, which is of the same type, and can therefore be used to create more objects from the factory. Note that we could even specify the inner class to be another `PayOffSpread`. The process has to end somewhere, since the number of cells used to make each successive `CellMatrix` gets smaller each time.

We still have to register these classes with the factory, this is done in `PayOffRegistration.cpp`

Listing 14.8 (`PayOffRegistration.cpp`)

```
#include <xlw/ArgListFactoryHelper.h>
#include "PayOffConcrete.h"

namespace
{
    FactoryHelper<PayOff,PayOffCall> callHelper("call");
    FactoryHelper<PayOff,PayOffPut> putHelper("put");
    FactoryHelper<PayOff,PayOffSpread> spreadHelper("spread");
}
```

Why do this is in a separate file? The reason is that if we decide to place the `PayOff` classes in a static library, then we cannot put the registrations in the library. The reason is that if we do, then they will be ignored! Material in a static library is only included when linking if it is referenced somewhere; a global variable declaration not mentioned anywhere will not be referenced and so not included.

14.11 Key points

In this chapter, we have seen how to create a templatized factory and met a few techniques along the way.

- Private inheritance can be used to express "implemented in terms of."
- The curiously recurring template pattern can be used to make the return type of a base class method equal to the type of the inherited class.
- The singleton can be implemented using the curiously recurring template pattern.
- We can use an argument list class to encapsulate a variable number of arguments of varying types.
- Some compilers have problems with the curiously recurring template pattern implementation of the singleton.
- The argument list class allows us to create a templatized factory without worrying about the types of arguments.
- The `CellMatrix` class gives us a way of transferring data to and from spreadsheets without having to deal with the particulars of the spreadsheet's data types.

14.12 Exercises

Exercise 14.1 Modify the random number classes to work with the `ArgumentList` factory. Include anti-thetic sampling and moment matching with arbitrary underlying classes amongst the classes to register.

Exercise 14.2 Modify the `xlw` factory so it uses the `Singleton` class developed here.

Exercise 14.3 Create a static library containing the pay-off classes from `xlw` and check how the registration works.

15

Interfacing with EXCEL

15.1 Introduction

The xlw package consists of a set of routines for building xlls. An xll is a dynamic link library (dll) that contains some special functions that allow the user to register functions with EXCEL. Once the xll has been created we simply open it from EXCEL, and some new functions appear that can be used just like ordinary inbuilt functions.

Our focus in this chapter is on how to use xlw. We do not address how it works. Indeed the philosophy of the current version is that using it should be similar to using a compiler – we wish to understand how to use all the features, but not how it works internally. The source code is fully available for those who are curious, however.

In this chapter, we will restrict our discussion to xlw 2.1. The package can be obtain from `xlw.sourceforge.net`. There is also an `xlw-users` mailing list which you can subscribe to for further discussion. The essential difference between the series 2 releases of xlw, which the author of this book wrote, and previous releases due to Jerome Lecomte and Ferdinando Ametrano is that the interfacing code is written automatically, so the user needs to know nothing about special data types or registration code.

The package comes with project files for 4 four different IDES: Visual Studio 6.0, 7.1, and 8.0, and for DevCpp. The DevCpp IDE is an open source IDE, which uses the MingW g++ compiler, so in particular this allows production of xlls using that compiler.

The xlw 2.1 package comes in three pieces:

- a console application run from the command line called `InterfaceGenerator`;
- a static library called xlwLib;
- and an example project with the name varying with compiler.

The user first has to build the `InterfaceGenerator` and xlwLib. Interfacing is done by applying `InterfaceGenerator` to header files at the command line. This then produces a C++ source file, which contains the code to interface the functions declared in that header file to EXCEL. A project then has to be built that links against xlwLib and includes the new source file.

The main trickinesses in the use of xlw are to do with how to set up projects and build for the first time; no actual interface coding is done by the user.

15.2 Usage

Before using xlw 2.1, we first have to build the xlw 2.1 library and the interface generator. The interface generator project can be found in the directory appropriate for your compiler:

- For DevCpp look in the folder xlwDevCpp, and the project is called InterfaceGenerator.dev.
- For Visual Studio 8.0, open the solution in the folder xlwVisio8, and the project is called InterfaceGenerator.
- For Visual Studio 7.1, open the solution in the folder xlwVisio7, and the project is called InterfaceGenerator.
- For Visual Studio 6.0, open the workspace in the folder xlwVisio6, and the project is called InterfaceGenerator.

This project should be built and will produce a console application called InterfaceGenerator.exe. Note that we can use the version of this application built with any one compiler with any other compiler without trouble.

Second we need to build the xlw 2.1 library to link against. The project files are in the same place as for the console application.

- For DevCpp the project file is called DevCppLibXl.dev, and the library file is called DevCppLibXl.a, and is built in to the same folder.
- For Visual Studio 8.0, the project is called xlwLib. The built libraries are xlwLib-Debug.lib and xlwLib.lib, and will be built into xlwLib/Release and xlwLib/Debug, respectively.
- For Visual Studio 7.1, the project is called xlwLib. The built libraries are xlwLib-Debug.lib and xlwLib.lib, and will be built into xlwLib7/Release and xlwLib7/Debug, respectively.
- For Visual Studio 6.0, the project is called xlwLib. The built libraries are xlwLib-Debug.lib and xlwLib.lib, and will be built into xlwLib6/Release and xlwLib6/Debug, respectively.

For each compiler, an example project is given of functions to be exported to the xll. These are called: DevCppXll.dev and xlwVisio. Each project contains a header file Test.h, a source file Test.cpp and an interface file xlwTest.cpp; these are contained in the folder TestFiles. Some files for payoffs can also be found there, and example spreadsheets.

It is the interface file xlwTest.cpp that has been automatically generated. To re-generate it, simply ensure that InterfaceGenerator.exe is in the path or in the same directory as Test.h and then at a command prompt type "InterfaceGenerator Test.h."

Simply building the xll project will then produce an xll, which can be opened in EXCEL and produces extra functions in a library called "MyTestLibrary."

To use xlw 2.1 for your own functions, you must first write a C++ function which compiles and builds except for the interfacing code. The functions to be exported to EXCEL should be contained in header files which contain nothing else. The InterfaceGenerator should then be applied to them. If the header file is called MyFile.h, the new file will be called xlwMyFile.cpp. The new file should then be added to the project. Note InterfaceGenerator will ignore any preprocessor commands, and will throw an error if the header file contains any classes or function definitions. It will also protest if any unknown data types are found; we discuss what data types are acceptable in Section 15.3.

The information for the function wizard in EXCEL is taken from comments and the names of argument variables in the header file. This means that arguments must be named. A comment should follow each argument name and this will appear in the function wizard when that argument is being entered. The general description of the function should be in a comment between the type of the function and its name.

Arguments can be passed by reference or by value, and can be const or non-const. (In fact, these have no effect on the coding of the interface file.)

Once the interface file has been added to the project, we simply build the project and then the output xll file should be openable by EXCEL. Note that we can have any number of interface files in the same xll project.

If you wish to create a new xll project, this can be done. The things to do are:

- The folder containing the xlw folder must be on the include path.
- The folder containing the xlw library file must be on the linking path.
- The xlw library file must be on the list of files to link against (i.e. for Visual studio, xlwLib.lib in release mode and xlwLib-Debug.lib in debug mode).
- The project must be a dll project in DevCpp. (Create a dll project, remove the file created by DevCpp, and then add your files.)
- The project must use multi-threaded dll code generation in Visual Studio. This means that you should create a dll project, or create a "Win32" application and then use "Application settings" to switch its type to a dll.

- Change the name of the output file to MyName.xll.

   Note if you are working with Visual Studio 8.0 Express, in addition, you must do the following:

- Install the Microsoft Platform SDK; this can be downloaded from the Microsoft website.
- The include directory for the SDK must be on the include path; this should happen automatically when you install the SDK.
- Link against the following libraries in debug mode: odbc32.lib odbccp32.lib, User32.lib, xlwLib-Debug.lib.
- Link against the following libraries in release mode: odbc32.lib, odbccp32.lib, User32.lib, xlwlib.lib
- Make sure the SDK library directory is included on the list of directories to search for library directories.
- When creating a new project, you must use create new project from existing code, and then later on say that it is a dll project. (This is not an option when creating new projects from new code.)

15.3 Basic data types

The function to be exported to EXCEL can only use data types supported by the interface generator. These are divisible into basic data types and extended types. The basic types are `double`, `short`, `NEMatrix`, `MyMatrix`, `MyArray`, `CellMatrix`, `string`, `std::string`, and `bool`. The extended data types are: `int`, `unsigned long`, `ArgumentList`, `DoubleOrNothing`, and `PayOff`.

The reason that `int` and `unsigned long` are extended types rather than basic types is that the type used by xlw to communicate with EXCEL is the `XLOPER`. This is a polymorphic data type with two numeric data types that are essentially `short` and `double`, so other numeric types go via double.

The class `MyMatrix` is defined via a typedef in `MyContainers.h` to `MJMatrix`. You can change this to your favourite matrix type. The matrix class must support the following: it should have `.rows()` and `.columns()` defined, a constructor that takes number of rows and columns, and elements should be accessible via `a[i][j]`. If your matrix class only supports element access via round brackets, you should define the macro `USE_PARENTHESESES`.

The class `NEMatrix` is a typedef for `MyMatrix`, but if you declare an argument to be of this type, then the function will not be called unless the argument is a non-empty matrix of numbers. (Otherwise, you get `#VALUE`.) If you are working with very large matrices, it should be more stable as the data type is much simpler and uses a different mechanism for transmitting the data to and from EXCEL. (For xll experts, it uses type "K" rather than type "P.")

The class `MyArray` is also defined via a typedef in `MyContainers.h`. The default is to `typedef` to `std::vector`. It must have `.size()`, a constructor taking the size, and `operator[]` defined.

We discussed the `CellMatrix` class at length in Section 14.8. The fact that this class allows a table of cells of arbitrary values including errors means that the conversion of EXCEL data to it should virtually never fail, since it allows error codes.

The types `std::string` and `string` are both allowed. These are the same class and the difference is simply in whether the namespace `std` has already been declared via `using`.

15.4 Extended data types

The xlw 2.1 package has been designed to make it easy to work with your own data types. The only constraint is that a function (or method) must exist that takes in a data type that is already constructible from basic types and creates the new type. We require the construction to be from a single previous type: argument specification would get rather complicated if multiple types were allowed. For this purpose, a constructor is equivalent to a function.

To add in extra types, we have to modify the `InterfaceGenerator` project. We simply add a declaration in the file `TypeRegistrations.cpp`. Note that the new classes themselves should not be included in the `InterfaceGenerator` project, since this project's role is to write C++ code and not to create executables.

For example

```
TypeRegistry::Helper
        arglistreg("ArgumentList", // new type
                    "CellMatrix", // old type
                    "ArgumentList", // converter name
                    false, // is a method
                    true, // takes identifier
                    "", // no key
                    "<xlw/ArgList.h>" // force inclusion
                                    // of this file
                        );


TypeRegistry::Helper
        payoffreg("Wrapper<PayOff>", // new type
                    "ArgumentList", // old type
```

```
"GetFromFactory<PayOff>",
 // converter name
false, // is a method
false, // takes identifier
"" , // no key
"<xlw/ArgListFactory.h>"
);
```

The first argument is the identifier for the new type.

The type to convert from is specified by the second argument.

The third is the function or method used to construct the new type from the old one.

The first bool is to specify whether the conversion function is a method of the old class, or simply a function or constructor that takes in an object of the old class.

The second bool indicates whether the converter method or function takes in a second argument that is a string expressing an identifier in case of error – this is very handy when trying to work out which argument in your complicated function is dubious.

For the curious only, the key is to tell EXCEL the type, this is generally only used when defining a basic type. This is typically "R" or "P." Doubles are passed as type "B" and non-empty matrices as "K." The types "R" and "P" indicate that the data are passed using the very useful but slightly painful data type XLOPER, which xlw then turns into an XlfOper. The type "K" means to pass using a floating point array, and "B" means pass directly as a double.

The last argument allows the forcing of extra #includes in our .cpp interface file. This allows us to ensure that the conversion function is available.

We can define new types from other new types. The maximum depth is 26, at which point the parser concludes that we have accidentally created a loop.

The three main data types that have been added for illustration are the DoubleOrNothing, ArgumentList and Wrapper<PayOff>. The ArgumentList we discussed at length in Chapter 14.

The DoubleOrNothing class allows us distinguish between a number passed in or an empty argument. We can therefore choose between a number passed in, and a default value if the argument is empty.

We illustrate using an argument list factory with EXCEL using the PayOff class. The factory returns a raw pointer to the base class, so this should be immediately converted to a smart pointer as we discussed in Section 13.3. Our new data type is therefore

```
Wrapper< PayOff >
```

which takes ownership and ensures deletion at the appropriate time. Note the point here that although the factory returns a raw pointer, the registration simply specifies the `Wrapper`, which silently takes ownership of the pointer.

We discuss briefly how this is implemented in the `InterfaceGenerator` project. The mechanism here is similar to that used for the factory. Every declaration of a `TypeRegistry::Helper` registers the new type with the `IncludeRegistry` class. This class is implemented using a singleton defined via the curiously recurring template pattern from Section 14.2.

15.5 xlw commands

When we look for your new functions in the function wizard in EXCEL, we will find that there is a new set of commands called "MyTestFile": the default name of the library in EXCEL is the name of the header file. We can change this by inserting the line

```
//<xlw:libraryname=MyTestLibrary
```

in the header file. Note that all functions in the header file will have the same library name, which will be that specified by the last `libraryname` command.

Some functions in EXCEL have the property of giving a different value each time they are called; for example, the time or the RAND() function. These functions are said to be `volatile`. If we want a `volatile` function, then this can be done as follows

```
double // system clock
//<xlw:volatile
SystemTime(DoubleOrNothing ticksPerSecond
// number to divide by
                    );
```

We can also time functions by inserting

```
//<xlw:time
```

in the same place. This causes two cells to be added below the function's results containing "time taken" and the time taken in seconds. It is possible to time `volatile` functions; the order of the two commands is not important.

15.6 The interface file

The interface generator creates the file `xlwMyTestFile.cpp`; it is not necessary to ever look at this interface file. However, it can be edited directly if so desired. It first has a `DummyFunction` declared in an unnamed namespace. This function references the functions `xlAutoOpen` and `xlAutoClose` and thus forces their inclusion in the xll. These functions carry out the registration of the other functions with EXCEL and are therefore essential.

There is a line

```
const char* LibraryName = "MyTestLibrary";
```

This specifies the name of the library in the EXCEL function wizard. This is enclosed in an anonymous namespace, so we can have multiple interface files in the same xll.

For each function, there are two parts. The first is the registration information. The second is the wrapper function called between EXCEL and the function chosen.

An example of the registration information is

```
namespace
{
XLRegistration::Arg
ConcatArgs[]=
{
{ "str1"," first string "},
{ "str2","second string "}
};

XLRegistration::XLFunctionRegistrationHelper
registerConcat("xlConcat",
"Concat",
" Concatenates two strings ",
LibraryName,
ConcatArgs,
"RR");
}
```

The code is placed in an unknown namespace to ensure it does not affect any linkage. The arguments are declared in the first part, with the name of each followed by its description.

In the second part, a global variable is declared. The creation of this global variable registers the function with a global singleton, which ensures that it is registered with EXCEL. This approach allows the registration to be split across many files. The information passed to the constructor is the name of the function in C++ in the interface file, the name of the function in EXCEL, the function description, the name of the library in EXCEL, the arguments declared above, and the types of the arguments. The types are expressed via a code, e.g.:

- R – LPXLOPER by reference
- P – LPXLOPER by value
- B – double
- K – floating point array

In xlw 2.1, only the types P, B, K, and R are used. P is used for `CellMatrix` and `MyMatrix`. K is used for `NEMatrix`. B for double. R is used for all other basic types. Types such as `bool` are therefore first passed in as LPXLOPERs and then transformed into the right data type.

An example of the interface function definition is

```
extern "C"
{
LPXLOPER EXCEL_EXPORT
xlConcat(
LPXLOPER xlstr1_,
LPXLOPER xlstr2_)
{
EXCEL_BEGIN;

if (XlfEXCEL::Instance().IsCalledByFuncWiz())
    return XlfOper(true);

XlfOper xlstr1(xlstr1_);
std::string str1(xlstr1.AsString("str1"));

XlfOper xlstr2(xlstr2_);
std::string str2(xlstr2.AsString("str2"));

std::string result(
    Concat(
            str1,
            str2)
    );
return XlfOper(result);
EXCEL_END
}
}
```

The extern 'C' command is necessary as we are using the C API and therefore must use C linkage.

The return type of the function is always LPXLOPER, but since this is a polymorphic data type, this is not a hindrance. EXCEL_EXPORT is a macro

```
#define EXCEL_EXPORT __declspec(dllexport)
```

This ensures that the function is a dll export, and so can be dynamically linked against.

Note that types are passed in as LPXLOPERs not XlfOpers or other types. XlfOpers are not used since they cause a crash with the MinGW compiler as they are not POD (plain old data) objects.

The macros EXCEL_BEGIN and EXCEL_END contain starting and finishing information common to all functions. In particular, EXCEL_END contains catches for common data types to return information to EXCEL.

The routine then checks if it is being called from the function wizard. If it is, then it immediately returns to EXCEL. This avoids time-consuming computations being called whilst data are being entered.

Each argument is then converted. First to an XlfOper, and then using the .As methods to the correct type. Note a string is passed into the .As method, this allows a throw to identify the offending argument. If you use extended types, there will be a string of conversions here.

Once the arguments have been converted, the original function is called and the result stored. For return to EXCEL, it is converted into an XlfOper and this is returned as an LPXLOPER.

15.7 The interface generator

The interface generator is written as a simple C++ routine. It is a console application that takes in one argument. The output file name is an optional argument and it defaults to adding "xlw" on the front and ".cpp" at the end.

The routine first reads in the file and places it in a vector of chars for convenience. It is then turned into tokens. Each token will be an identifier, preprocessor directive, comment, or delimiter.

Tokens corresponding to consts and ampersands are then removed, since they will not affect the coding of the interface routine. At this stage, unsigned identifiers are also dealt with.

Once this has been done, the file is turned into a list of functions, with each function having a list of arguments with names and types.

The next operation is to identify all the types and find the conversion routines.

Once this has been done, the output file is written into a vector, and then written to a file.

15.8 Troubleshooting

If you find that you can build the xll but that the functions do not register, here are some common problems with xlls in general.

If absolutely nothing happens, then check security settings for macros. The default setting is to ignore files containing macros. Note for a reasonable level of security setting, you will get asked whether to enable macros; if this does not happen then security levels are too high or possibly too low.

If you get an error saying that the file is in an "unrecognizable format," possibilities are

- Missing dlls on your machine. This often occurs if you compile in VC7 on one machine and then move the xll to another one. This can be cured either by working out how to get the compiler not to need the dlls, or by copying the required dlls to the new machine (or switch to using DevCpp.) Another solution is to download "Microsoft Visual C++ 2005 Redistributable Package (x86)" which is designed to solve this sort of problem but needs to be installed on the new machine.
- Failure to export the auto open and auto close functions.

The dumpbin utility can be used to check whether the right functions are being exported. If you have followed the instructions here, this *should* not be an issue.

15.9 Debugging with xlls

Suppose you have managed to build and run the xll. Now you want to debug it. It is still possible to use the Visual Studio debugger. Here are the precise instructions for Visual C++ 8.0. The approach is similar for the other Visual Studios. DevCpp does have an inbuilt debugger, GDB, but it seems painful to use it.

- In the project properties, go to the "Debugging" part of the "Configuration" properties . In the "Command" entry select "Browse" and then browse for the EXCEL executable.
- Below this, look for "command arguments." In here we want the xll. To get this go to the "Linker" part of "Configuration" properties and look for the value of "output file." Copy this and paste it into "command arguments."
- Hit the F5 key.

The debugger should now stop on breakpoints in your C++ code, as usual, when your functions are used.

You can now exit in two ways: either by using the "stop debugging" command in the "debug" menu, or by exiting EXCEL in the ordinary way. The first method causes EXCEL to think that it crashed, and it will ask you about file recovery on next being run. It is therefore best avoided.

15.10 Key points

- Quants generally don't use console applications.
- xlls are a standard way of interfacing with EXCEL.
- xlw gives an easy way to create xlls.
- All interfacing code is generated automatically by xlw.

15.11 Exercises

Exercise 15.1 Download and build xlw.

Exercise 15.2 Create an xll that prices a Black–Scholes option using the analytic formula from the appendix.

Exercise 15.3 Set up a new xlw project.

Exercise 15.4 Modify the InterfaceGenerator project so that it interfaces random number generators.

16

Decoupling

16.1 Introduction

Have you ever worked on a library that you were scared to rebuild because it would take too long? Or have you been afraid to change the contents of a file because of the knock-on effects making everything recompile? Have you ever included a file in your project and discovered that this made you include a large number of seemingly irrelevant files in your project to get rid of unresolved symbol linker errors?

We have focussed in this book on issues relating to logical design and code reuse. These examples are, however, not problems with logical design; for example, adding an extra data member to a class forces every client to recompile but will not cause any of them to have compilation errors. All these are examples of *physical design* problems. Encapsulation, which we have discussed and illustrated at length, assists with logical design but not physical design.

Physical design problems indicate pieces of code that are more tightly coupled than they should be. In this chapter, we study physical design practice, and, in particular, examine the concept of *insulation*. This is a much stronger requirement than encapsulation, the difference being that if class A is insulated from class B, then changes to A do not cause B to recompile. Whereas encapsulation merely guarantees that B will not have to be recoded.

16.2 Header files

The fundamental difference between header files and source files is that header files are included by other files both header and source via #include, whereas source files should not be. The first consequence of this is that every time you use #include in a header file, you are including a file not just into that header file but also into every file that includes it both directly and indirectly.

Unnecessary header file inclusions therefore slow compile times in two ways, first by causing the compiler to look at much more code for each compiled source file, and second by causing more source files to require compilations. How can we reduce the numbers of #includes?

The first and most important method is simply by being aware of the issue. If you are ever tempted to #include a file just because your client might need it, don't! If your client needs it, he will include it for himself.

A second method which can make a large difference is to use forward declarations. Often in a header file we simply need to know that a class exists, we don't need to know anything more about it. For example, suppose we have declared a function f in our header file that takes in a CellMatrix and returns a double. We could code this in two ways

Listing 16.1 (ExampleFile1.h)

```
#ifndef EXAMPLE_FILE1_H
#define EXAMPLE_FILE1_H

#include "CellMatrix.h"

double f(const CellMatrix& cells);

#endif
```

or by

Listing 16.2 (ExampleFile2.h)

```
#ifndef EXAMPLE_FILE2_H
#define EXAMPLE_FILE2_H

class CellMatrix;

double f(const CellMatrix& cells);

#endif
```

The first method forces any file that includes ExampleFile1.h to recompile every time a change is made to CellMatrix.h or to a file included by it. The second does not. Here is a non-trivial example of this technique taken from xlw:

```
#ifndef DOUBLE_OR_NOTHING_H
#define DOUBLE_OR_NOTHING_H
class CellMatrix;
#include <xlw/port.h>
#include <string>
```

```
class DoubleOrNothing
{
public:
        DoubleOrNothing(const CellMatrix& cells,
                        const std::string& identifier);

        bool IsEmpty() const;
        double GetValueOrDefault(double defaultValue) const;

private:
        bool Empty;
        double Value;

};


#endif
```

The DoubleOrNothing class uses the CellMatrix in its constructor only so it does not have to include the file from its header file. Of course, the source file does include it

Listing 16.3 (DoubleOrNothing.cpp)

```
#include "DoubleOrNothing.h"
#include "CellMatrix.h"

DoubleOrNothing::DoubleOrNothing(const CellMatrix& cells,
                                 const std::string& identifier)
{
   if (cells.ColumnsInStructure() != 1
       || cells.RowsInStructure() != 1)
      throw("Multiple values given "
            "where one expected for DoubleOrNothing "
                +identifier);


   if (!cells(0,0).IsEmpty() && !cells(0,0).IsANumber() )
      throw("expected a double or nothing, got something else "
            +identifier);

   Empty = cells(0,0).IsEmpty();
```

```
    Value = Empty ? 0.0 : cells(0,0).NumericValue();

}


bool DoubleOrNothing::IsEmpty() const
{
        return Empty;
}


double DoubleOrNothing::GetValueOrDefault(double defaultValue)
                            const
{
        return Empty ? defaultValue : Value;
}
```

If you get the error "undefined class" when compiling, this means that you have for-ward declared a `class` but not included it when it is really needed. Note that clients passed a `DoubleOrNothing` generally will not need to include `CellMatrix.h`, since this is only needed for the constructor.

When can we use this technique? As long as we have no variables of the type of the class, and we attempt to call no methods of the class, then a forward declaration is sufficient. This means that if we have data members of type `CellMatrix`, then we must include `CellMatrix.h`, but we can have pointers of type `CellMatrix*` if we wish without inclusion. We can also use forward-declared classes as arguments of functions and as return types.

One thing we cannot do is forward declare identifiers that are `typedef`ed. So if we have defined `MyArray` by `typedef`, then we must always include the file with the `typedef`. Note the unpleasant consequence of this is that if we have forward declared our array class everywhere and then decide that we want to switch to an-other class, then we will get compilation errors everywhere if we decide to achieve the change via `typedef`.

16.3 Splitting files

An important physical design issue is how to divide up code. The simplest form of this question is "how many classes do I put in one file?" An easy answer is one. The only problem with this is that you will find that you very quickly have hundreds of files, which you may or may not find annoying.

An important rule to follow is that abstract base classes should have their own files. Most clients will only ever need to see the declaration of the base class, since they will work with references to the base class or (smart) pointers to base class objects. Since most base classes have no data members, this can mean no implied inclusions.

This effect is multiplied when we use factories – in that case, even the file that instantiates the object of type X never needs to include the class declaration of X, and, in fact, it never even needs to know that X exists. For an example of this approach, see the files `PayOff.h` and `PayOffConcrete.h` in the xlw project. The inherited class, which are *concrete* rather than abstract, are in `PayOffConcrete.h`, but `PayOff.h` is almost empty but it is all that clients need to see

Listing 16.4 (`PayOff.h`)

```
#ifndef PAYOFF_H
#define PAYOFF_H
class PayOff
{
public:

    PayOff();

    virtual double operator()(double Spot) const=0;
    virtual ~PayOff();
    virtual PayOff* clone() const=0;

private:

};
#endif
```

In addition, the files `Test.h` and `Test.cpp`, which use the `PayOff` classes, never include `PayOffConcrete.h`. Try this: make a small change to `PayOffConcrete.h`, and hit build; the only files that recompile are `PayOffConcrete.cpp` and `PayOffRegistration.cpp`.

A secondary rule worth following is to split files when classes require quite different inclusions. The inclusions required give a good indicator of similarity and they are the aspect of the class that has the biggest knock-on effect.

16.4 Direction of information flow and levelization

When implementing two related classes X and Y, we need to think about what we really need to know about the other. For example, consider the issue of interfacing with EXCEL. We want to be able to input matrices of type `MyMatrix` from

EXCEL; we also want to be able to return them to EXCEL via the `XlfOper` data type.

How can we implement this functionality? There are a number of solutions:

(i) Give `MyMatrix` a constructor that takes in an `XlfOper` and a method `.AsXlfOper()` that turns it into an `XlfOper`.
(ii) Give `XlfOper` a constructor that takes in a `MyMatrix` and a method `.AsMatrix()` that converts back.
(iii) Do both (i) and (ii).
(iv) Write functions

```
MyMatrix ConvertToMatrix( const XlfOper& inputOper);
XlfOper ConvertToXlfOper(const MyMatrix& inputMatrix);
```

and put them in a file on their own.

If we adopt solution (i), then every time we use the `MyMatrix` class we must include the `XlfOper` class and everything that `XlfOper` requires; this is true in both compile and link senses. However, we can create matrices and convert them to `XlfOper`s with little coding. Whereas if we adopt solution (ii), then we have the opposite dependency, and conversion back and forwards is still easy to write.

Solution (iii) means that we have dependencies both ways, and choices on how to do the conversion. Choice is not necessarily good, since it always raises the question of which way to do it.

Solution (iv) creates no dependencies between the two classes. We can use matrices without worrying about EXCEL interfacing and vice versa. The main downside is that it is a little clunky syntactically.

In order to decide which is best, we need to consider what direction information flows in, and what dependencies we are happy with. Solution (i) means all our numerical code is dependent on our EXCEL interfacing code. If we decide that we want to use a different spreadsheet package, we are stuck with including the xlw files. This is clearly a bad idea; this rules out solutions (i) and consequently (iii) as well.

Solution (ii) has the opposite dependency: the xlw code has to know about the matrix class. This is not so bad, since if we change the matrix class, we will still want to be able to interface the new class with EXCEL; changing the matrix class will force recompilation of some of the xlw files but not have other drastic effects. It is very unlikely that we will ever want to get rid of matrices altogether.

We have already ruled out solution (iii), but we discuss the consequences a little further. Suppose we had gone down this route. This would mean that it was impossible to use the matrix class without the EXCEL interfacing code, and it would be impossible to use the EXCEL interfacing code without the matrix class. At that point, we might as well put them in the same file, since each one requires the other.

Solution (iv) is nice in that we have fully decoupled the two classes; we simply have an extra file that depends on both of them. On the other hand, the syntax is a little non-obvious, and we will spend a certain amount of time having to look it up each time we do the conversion.

The xlw package uses solution (ii) on the grounds that the extra decoupling does not buy us anything in this particular case and, given the natural dependencies between classes, the additional dependency is not an issue. Note that the entire package does this for all conversions.

In general, think about which direction information flows naturally, and whether you really want users of class X to be forced to deal with Y and vice versa. If in doubt, use solution (iv), since this gives the maximum amount of decoupling.

Another way to think about this is in terms of fundamental types: a type should only depend on types that are more fundamental than it. So in xlw, the `PayOff` classes depend on `ArgumentLists`, which depend on the `CellMatrix` class but not the other way round. This means that we should avoid solution (iii) as much as possible. If it seems that two classes are at a similar level of fundamentalism, then you should probably be using solution (iv).

This leads to the idea of *levelization*. We assign a level number to each class; the more fundamental a class is, the lower its level is. A class or file should only depend on classes and files that are on strictly lower levels. We should therefore never have a pair of classes that both depend on each other, either directly or indirectly.

16.5 Classes as insulators

Consider the `CellMatrix` class, this was introduced to insulate EXCEL from the numerical code. It encapsulates the notion of an input table of cells from *any* spreadsheet. It can therefore carry any information that can be input from a spreadsheet, which can then be converted into any class as desired via argument lists.

However, the `CellMatrix` class itself does not depend in any way on the code for interfacing with EXCEL. It therefore acts as a messenger between EXCEL and the numerical code without creating dependencies: both the numerical code and the EXCEL code depend on `CellMatrix`, but it depends on neither.

In fact, the `CellMatrix` class depends only on basic container classes and the standard library. We can therefore view it as being on a very low level. Note the general technique here: we can achieve decoupling by using a low-level class to communicate between disparate pieces of code.

16.6 inlining

The keyword `inline` is often used for optimization. What does it do? It gives the compiler the *option* of replacing a function call with the actual code defining the function. This eliminates a function call, thus saving a little time. In addition, the fact that the code is being considered at the same time as the surrounding code allows an optimizing compiler to make additional optimizations.

A typical use for `inline` is for data access operations. For example, when defining an array class, it is common to make `operator[]` an `inline` method. Thus encapsulation of `private` data is preserved with no speed cost at run-time.

The first point to be aware of is that the `inline` only gives the compiler an option; it can be happily ignore the suggestion if it wants to.

A second more important point is that whilst we have preserved encapsulation at no run-time cost, we have lost a lot of insulation at plenty of compile- and link-time cost. For `inline` to work, the function must be provided in the header file; thus we now have dependency on the implementation of the function or method. If we change the function's implementation, all clients must recompile. With a non-`inline` function defined in the source file, the impact would be minimal. Note also that defining a function in a header file may also require extra `#includes`, creating further dependencies.

Another aspect of inlining to consider is that if the function is non-trivial and heavily used, there will be a lot of copies of the function sprinkled throughout the code, which will increase executable size.

In conclusion, whilst inlining is useful for avoiding performance overheads with `private` data, be aware of the downsides and do not overuse it.

16.7 Template code

The standard way to work with template code is to first declare the template class or function in a header file, and then in the same header file provide the method or function definitions.

We therefore have no insulation between the class and its clients. Any change to any aspect of a template class will force recompilation of all its clients. This is quite different from non-template code, where changing the implementation of functions and methods has no compile-time effects on clients.

The gains and losses of template code are really very similar to those with `inline`. We gain on run-time speed and reusability, but lose on compile time.

The reader is probably wondering why template code has to be done this way. Why can't we have template code defined in a source file? The answer is that we can, but that most modern compilers are not yet up to it. The keyword for doing this is `export` and only one compiler, the Comeau compiler, currently supports it. It is sufficiently controversial that there were motions to remove it from the next C++ standard; however, these were defeated so `export` will remain in the standard at least, for a while.

16.8 Functional interfaces

When we discussed the conversion between matrices and XlfOpers, one solution was to simply use two functions contained in a separate file. This can be viewed as a functional interface approach to decoupling. By introducing functions in a third file we can convert between disparate objects without them knowing anything about each other.

Another example of functional interface decoupling is the EXCEL C API. Communicating via a simple functional gateway with C linkage allows both sides of the gate to change dramatically without forcing changes on the other side. If we build an xll for one version of EXCEL and then up-grade to a new version, then the xll will still work. Similarly, EXCEL knows nothing about the interiors of our xlls.

The key here is that two quite different components have had their method of communication restricted to occur in a quite precise way. As long as this remains true, the two pieces can vary without knock-on effects.

16.9 Pimpls

In this section, I want to briefly mention an idiom that I do not use personally; however, it represents the ideas in this chapter pushed as far as they will go, so the reader should be aware of it.

The essential difference between encapsulation and insulation is that when we change the `private` section of the class then we do not affect classes that are protected by encapsulation, but we do fail the insulation test because all client classes have to recompile.

If we could move all the `private` section into the source file, then this problem would disappear. The PIMPL idiom is one way to achieve this. PIMPL stands for "Private Implementation." The essence of the approach is to have the class that is visible to clients, A, contain no data except a pointer to an undefined class, B, which is defined in the source file. In addition, we place B in an unnamed `namespace` so it has internal linkage and can have no compile- or link-time effects on other files.

It is B that contains all the data, so when we change the class data members there is no knock-on effects. We have thus achieved perfect insulation.

What are the consequences of this approach? The first point to note is that the data member of A is a pointer, so all memory allocation has to be dynamic; that is, we must use `new` to create objects of type B and then manage their copying and destruction as well. Whilst the coding pain can be minimized by using an appropriate smart pointer, it is still some extra fiddliness.

We also have the issue that `new` is slow. This means that we really do not want to do this when speed is important. This not a huge issue in that it is rarely wise to create and destroy objects of any non-trivial class in the core part of numerical routines where all the time is taken, in any case; this criticism equally applies to any class that has a container as a data member.

There is also a memory cost; we need an extra pointer for each object from the class. This is not a real problem; if you are using many objects from this class and the size of a pointer is noticeable, you should probably not be using this idiom for other reasons such as speed.

The main downside for me with PIMPL is that the additional clunkiness outweighs the gains in insulation. It is therefore not a great idea for numerical coding. However, if you ever need to work on a large system that has heavy amounts of non-numerical code, it is worth considering. See [32] for an enthusiastic exposition of the technique, and also for some discussion of how to avoid the time consumption issues with memory allocation.

16.10 Key points

In this chapter, we have examined how to reduce, compile, and link dependencies using the concept of insulation.

- Physical design relates to how files depend on each other.
- Encapsulation stops us from having to rewrite code.
- Insulation saves us from having to recompile code.
- Being careful with #include can speed up compile times.
- Levelization yields a natural way to eliminate excess dependencies.
- Inlining has costs in terms of physical design as well as gains in terms of run speed.
- Template code increases coupling.
- PIMPL is a powerful methodology for minimizing dependencies.

This chapter has been greatly influenced by Lakos's book, *Large Scale C++ Software Design,* [16]. Whilst it is now a little dated, there is still a wealth of useful discussion and techniques. Sutter's "Exceptional" books, [32, 33, 34] also discuss similar issues at length, but are more recent.

16.11 Exercises

Exercise 16.1 Take the most recent project you have completed and time a "rebuild all." Now see how many #includes you can eliminate via forward declaration and time again. Repeat after splitting out abstract base classes.

Exercise 16.2 Implement a class using the PIMPL idiom.

Appendix A

Black–Scholes formulas

In Chapter 9, we developed an implied volatility function; this function necessarily meant we needed formulas for the Black–Scholes price and vega of a call option. We present code for the pricing of calls, puts, digital calls and digital puts, and for the vega of a call option in BlackScholesFormulas.h and BlackScholes-Formulas.cpp. We do not present the formulas here as they can be found in just about any book on derivatives pricing, e.g. [13]. The code is a straightforward implementation of formulas. Note that we rely heavily on having an implementation of the cumulative normal function. We give such an implementation in Appendix B.

Listing A.1 (BlackScholesFormulas.h)

```
#ifndef BLACK_SCHOLES_FORMULAS_H
#define BLACK_SCHOLES_FORMULAS_H

double BlackScholesCall( double Spot,
                         double Strike,
                         double r,
                         double d,
                         double Vol,
                         double Expiry);

double BlackScholesPut( double Spot,
                        double Strike,
                        double r,
                        double d,
                        double Vol,
                        double Expiry);
```

```
double BlackScholesDigitalCall(double Spot,
                               double Strike,
                               double r,
                               double d,
                               double Vol,
                               double Expiry);


double BlackScholesCallVega( double Spot,
                             double Strike,
                             double r,
                             double d,
                             double Vol,
                             double Expiry);

#endif
```

Listing A.2 (BlackScholesFormulas.cpp)

```
#include <BlackScholesFormulas.h>
#include <Normals.h>
#include <cmath>

#if !defined(_MSC_VER)
using namespace std;
#endif

double BlackScholesCall( double Spot,
                         double Strike,
                         double r,
                         double d,
                         double Vol,
                         double Expiry)
{
  double standardDeviation = Vol*sqrt(Expiry);
  double moneyness = log(Spot/Strike);
  double d1 =( moneyness + (r-d)*Expiry +
  0.5* standardDeviation*standardDeviation)/standardDeviation;
  double d2 = d1 - standardDeviation;
  return Spot*exp(-d*Expiry) * CumulativeNormal(d1) -
   Strike*exp(-r*Expiry)*CumulativeNormal(d2);
}
```

```
double BlackScholesPut( double Spot,
                        double Strike,
                        double r,
                        double d,
                        double Vol,
                        double Expiry)
{
  double standardDeviation = Vol*sqrt(Expiry);
  double moneyness = log(Spot/Strike);
  double d1 =( moneyness + (r-d)*Expiry +
  0.5* standardDeviation*standardDeviation)/standardDeviation;
  double d2 = d1 - standardDeviation;
  return Strike*exp(-r*Expiry)*(1.0-CumulativeNormal(d2)) -
   Spot*exp(-d*Expiry) * (1-CumulativeNormal(d1));
}


double BlackScholesDigitalCall( double Spot,
                                double Strike,
                                double r,
                                double d,
                                double Vol,
                                double Expiry)
{
  double standardDeviation = Vol*sqrt(Expiry);
  double moneyness = log(Spot/Strike);
  double d2 =( moneyness + (r-d)*Expiry -
  0.5* standardDeviation*standardDeviation)/standardDeviation;
  return exp(-r*Expiry)*CumulativeNormal(d2);
}


double BlackScholesDigitalPut( double Spot,
                               double Strike,
                               double r,
                               double d,
                               double Vol,
                               double Expiry)
{
  double standardDeviation = Vol*sqrt(Expiry);
  double moneyness = log(Spot/Strike);
  double d2 =( moneyness + (r-d)*Expiry -
```

```
    0.5* standardDeviation*standardDeviation)/standardDeviation;
    return exp(-r*Expiry)*(1.0-CumulativeNormal(d2));
}


double BlackScholesCallVega( double Spot,
                             double Strike,
                             double r,
                             double d,
                             double Vol,
                             double Expiry)
{
    double standardDeviation = Vol*sqrt(Expiry);
    double moneyness = log(Spot/Strike);
    double d1 =( moneyness +  (r-d)*Expiry +
    0.5* standardDeviation*standardDeviation)/standardDeviation;
    return Spot*exp(-d*Expiry) * sqrt(Expiry)*NormalDensity(d1);
}
```

Appendix B

Distribution functions

We have repeatedly used the cumulative normal distribution function for a standard normal random variable and its inverse. We give an implementation of rational approximations for these functions in `Normals.h` and `Normals.cpp`. For further discussion of these approximations see [5], [21] and [10].

Listing B.1 (`Normals.h`)

```
#ifndef NORMALS_H
#define NORMALS_H


double NormalDensity(double x);


double CumulativeNormal(double x);


double InverseCumulativeNormal(double x);


#endif
```

Listing B.2 (`Normals.cpp`)

```
/*
code to implement the basic distribution functions necessary
inmathematical finance via rational approximations
*/


#include <cmath>
#include <Normals.h>


// the basic math functions should be in namespace std but
// aren't in VCPP6
```

```
#if !defined(_MSC_VER)
using namespace std;
#endif


const double OneOverRootTwoPi = 0.398942280401433;


// probability density for a standard Gaussian distribution
double NormalDensity(double x)
{
    return OneOverRootTwoPi*exp(-x*x/2);
}


// the InverseCumulativeNormal function via the
// Beasley-Springer/Moro approximation
double InverseCumulativeNormal(double u)
{

    static double a[4]={  2.50662823884,
                        -18.61500062529,
                         41.39119773534,
                        -25.44106049637};

    static double b[4]={-8.47351093090,
                        23.08336743743,
                       -21.06224101826,
                        3.13082909833};

    static double c[9]={0.3374754822726147,
                        0.9761690190917186,
                        0.1607979714918209,
                        0.0276438810333863,
                        0.0038405729373609,
                        0.0003951896511919,
                        0.0000321767881768,
                        0.0000002888167364,
                        0.0000003960315187};

    double x=u-0.5;
    double r;
```

```
    if (fabs(x)<0.42) // Beasley-Springer
    {
        double y=x*x;

        r=x*(((a[3]*y+a[2])*y+a[1])*y+a[0])/
                ((((b[3]*y+b[2])*y+b[1])*y+b[0])*y+1.0);

    }
    else // Moro
    {

        r=u;

        if (x>0.0)
            r=1.0-u;

        r=log(-log(r));

        r=c[0]+r*(c[1]+r*(c[2]+r*(c[3]+r*(c[4]+r*(c[5]+
                r*(c[6]+r*(c[7]+r*c[8])))))));

        if (x<0.0)
            r=-r;

    }

    return r;
}


// standard normal cumulative distribution function
double CumulativeNormal(double x)
{
    static double a[5] = { 0.319381530,
                          -0.356563782,
                           1.781477937,
                          -1.821255978,
                           1.330274429};

    double result;
```

```
if (x<-7.0)
    result = NormalDensity(x)/sqrt(1.+x*x);

else
{
    if (x>7.0)
        result = 1.0 - CumulativeNormal(-x);
    else
    {
        double tmp = 1.0/(1.0+0.2316419*fabs(x));

        result=1-NormalDensity(x)*
                (tmp*(a[0]+tmp*(a[1]+tmp*(a[2]+
                tmp*(a[3]+tmp*a[4])))));

        if (x<=0.0)
            result=1.0-result;
    }
}
return result;
}
```

Appendix C

A simple array class

C.1 Choosing an array class

To do any serious numerical work, we need the concept of an array. In C++ this translates into the need for an array class. There are any number of array classes available on the web, and the coder can easily write his own. Indeed most learners of C++ develop their own array class as an exercise sometime early in their study. In fact, the C++ standard library includes an array class which is designed to be optimized for high-speed numerical computation. This template array class is called `valarray` and has different emphases from the standard library's `vector` class. Where `vector` was designed to be efficient for resizing and inserting elements, the assumption in `valarray` is that the array's size will not often vary. Instead `valarray` is designed for speed, and it provides many numerical operations. For a discussion of the `valarray` class see [12] or [31].

There is, however, a downside to using `valarray`: there is no range checking. So if your index goes off the end of the array, you do not receive a nice error message, but instead you just get garbage or a crash. The reason that range checking is not provided is that it is time consuming, and logically correct code will, by definition, not need it. However, it is an extremely useful debugging tool, and so ideally it should be easy to switch on and off.

Our solution is therefore to provide our own array class which is modelled on `valarray<double>` and provides a subset of its operations. This means that we can use a `typedef` to replace our class with `valarray<double>` when we are confident in our code. This allows us to take advantage of all the optimizations in `valarray`, whilst retaining an easy-to-use class for debugging. Our class provides range-checking if and only if the macro RANGE_CHECKING is defined. Thus by changing one project setting, we can quickly shift between safe and fast modes. We also provide the facility to shift between `valarray<double>` and our class via defining the macro USE_VAL_ARRAY.

It is important to be careful when choosing an array class. Once a class interface has been chosen, you are stuck with it. You can always gut the insides of your class to make it more efficient but once the class has been used throughout your code, a lot of time has been invested in the interface, and to change it is a difficult proposition.

C.2 The header file

We present the class, called `MJArray` to avoid the likelihood of name clashes, in `Arrays.h`.

Listing C.1 (`Arrays.h`)

```
#ifndef MJARRAYS_H
#define MJARRAYS_H

#ifdef USE_VAL_ARRAY

#include <valarray>

typedef  std::valarray<double> MJArray;

#else   // ifdef USE_VAL_ARRAY

class MJArray
{

public:
    explicit MJArray(unsigned long size=0);
    MJArray(const MJArray& original);

    ~MJArray();

    MJArray& operator=(const MJArray& original);
    MJArray& operator=(const double& val);

    MJArray& operator+=(const MJArray& operand);
    MJArray& operator-=(const MJArray& operand);
    MJArray& operator/=(const MJArray& operand);
    MJArray& operator*=(const MJArray& operand);
```

```
    MJArray& operator+=(const double& operand);
    MJArray& operator-=(const double& operand);
    MJArray& operator/=(const double& operand);
    MJArray& operator*=(const double& operand);

    MJArray apply(double f(double)) const;

    inline double operator[](unsigned long i) const;
    inline double& operator[](unsigned long i);

    inline unsigned long size() const;

    void resize(unsigned long newSize);

    double sum() const;
    double min() const;
    double max() const;

private:
    double* ValuesPtr;
    double* EndPtr;

    unsigned long Size;
    unsigned long Capacity;
};

inline double MJArray::operator[](unsigned long i) const
{
#ifdef RANGE_CHECKING
    if (i >= Size)
    {
        throw("Index out of bounds");
    }
#endif

    return ValuesPtr[i];
}

inline double& MJArray::operator[](unsigned long i)
{
#ifdef  RANGE_CHECKING
```

```
    if (i >= Size)
    {
        throw("Index out of bounds");
    }
#endif

    return ValuesPtr[i];
}

inline unsigned long MJArray::size() const
{
    return Size;
}
#endif // ifdef USE_VAL_ARRAY
#endif // ifndef MJARRAYS_H
```

We have provided the standard class operations: constructor, copy constructor, assignment operator, and destructor. We can carry out simple numerical operations, such as addition or multiplication, both by `doubles`, which are applied to each element, and by arrays, which are applied pointwise.

We have overloaded `operator[]`. It is provided in `const` and non-`const` versions. The former can be used to read elements of `const` arrays, whereas the latter can be used to modify elements of non-`const` arrays. Note that these operators have been inlined. This means that when the compiler encounters the operator, it reproduces the code inside the function instead of setting up a call to the function. This ensures that no extra overhead is caused by going via the class interface. (An alternative would have been to make the data pointer public and use C style array access, but this would have badly violated one of our basic rules that the class data members should be private.)

We also include the `apply` method. This takes in a function object, `f`, and applies it to each element of the array and outputs a new array consisting of the results.

We can take the `size` of our array, and we can resize it. We follow the `valarray` class in not requiring the resize operation to preserve the underlying data.

Finally, we also include the self-explanatory operations `sum`, `min`, and `max`. Valarray contains many more operations which could be added to our class as and when necessary.

Our implementation has four data members. We have pointers to express the beginning and end of the array. We also have two different size members. The member `Size` expresses the number of elements currently in the array, whereas the member `Capacity` expresses the amount of memory that has been allocated. So we should always have that `Size` is less than or equal to `Capacity`, and that `Size` is equal to `EndPtr` minus `StartPtr`.

C.3 The source code

We present the source code for our array class in `Arrays.cpp`.

Listing C.2 (`Arrays.cpp`)

```cpp
#include <Arrays.h>
#include<algorithm>
#include<numeric>

MJArray::MJArray(unsigned long size)
: Size(size), Capacity(size)
{
    if (Size >0)
    {
        ValuesPtr = new double[size];
        EndPtr = ValuesPtr;
        EndPtr += size;
    }
    else
    {
        ValuesPtr=0;
     EndPtr=0;
    }
}


MJArray::MJArray(const MJArray& original)
:
    Size(original.Size), Capacity(original.Size)
{
    if (Size > 0)
    {
        ValuesPtr = new double[Size];

        EndPtr = ValuesPtr;

        EndPtr += Size;

        std::copy(original.ValuesPtr, original.EndPtr, ValuesPtr);
```

```
    }
    else
    {
        ValuesPtr = EndPtr =0;
    }
}

MJArray::~MJArray()
{
    if (ValuesPtr >0)
        delete [] ValuesPtr;
}

MJArray& MJArray::operator=(const MJArray& original)
{
    if (&original == this)
    return *this;

    if (original.Size > Capacity)
    {
        if (Capacity > 0)
            delete [] ValuesPtr;

            ValuesPtr = new double[original.Size];

            Capacity = original.Size;
    }

    Size=original.Size;

    EndPtr = ValuesPtr;
    EndPtr += Size;

    std::copy(original.ValuesPtr, original.EndPtr, ValuesPtr);

    return *this;
}

    void MJArray::resize(unsigned long newSize)
    {
    if (newSize > Capacity)
```

```
    {
        if (Capacity > 0)
            delete [] ValuesPtr;

        ValuesPtr = new double[newSize];

        Capacity = newSize;
    }
    Size = newSize;

    EndPtr = ValuesPtr + Size;
}

MJArray& MJArray::operator+=(const MJArray& operand)
{
#ifdef RANGE_CHECKING
    if ( Size != operand.size())
    {
        throw("to apply += two arrays must be of same size");
    }
#endif

    for (unsigned long i =0; i < Size; i++)
        ValuesPtr[i]+=operand[i];

    return *this;
}
MJArray& MJArray::operator-=(const MJArray& operand)
{
#ifdef RANGE_CHECKING
    if ( Size != operand.size())
    {
        throw("to apply -= two arrays must be of same size");
    }
#endif

    for (unsigned long i =0; i < Size; i++)
        ValuesPtr[i]-=operand[i];

    return *this;
}
```

```
MJArray& MJArray::operator/=(const MJArray& operand)
{
#ifdef RANGE_CHECKING
    if ( Size != operand.size())
    {
        throw("to apply /= two arrays must be of same size");
    }
#endif

    for (unsigned long i =0; i < Size; i++)
        ValuesPtr[i]/=operand[i];

    return *this;
}


MJArray& MJArray::operator*=(const MJArray& operand)
{
#ifdef RANGE_CHECKING
    if ( Size != operand.size())
    {
        throw("to apply *= two arrays must be of same size");
    }
#endif

    for (unsigned long i =0; i < Size; i++)
        ValuesPtr[i]*=operand[i];

    return *this;
}


/////////////////////////////

MJArray& MJArray::operator+=(const double& operand)
{
    for (unsigned long i =0; i < Size; i++)
        ValuesPtr[i]+=operand;

    return *this;
}


MJArray& MJArray::operator-=(const double& operand)
```

```
{
    for (unsigned long i =0; i < Size; i++)
        ValuesPtr[i]-=operand;

    return *this;
}

MJArray& MJArray::operator/=(const double& operand)
{
    for (unsigned long i =0; i < Size; i++)
        ValuesPtr[i]/=operand;

    return *this;
}

MJArray& MJArray::operator*=(const double& operand)
{
    for (unsigned long i =0; i < Size; i++)
    ValuesPtr[i]*=operand;

    return *this;
}

MJArray& MJArray::operator=(const double& val)
{
    for (unsigned long i =0; i < Size; i++)
        ValuesPtr[i]=val;

    return *this;
}

double MJArray::sum() const
{
    return std::accumulate(ValuesPtr,EndPtr,0.0);
}

double MJArray::min() const
{
#ifdef RANGE_CHECKING
    if ( Size==0)
    {
```

```
            throw("cannot take min of empty array");
        }
#endif RANGE_CHECKING
        double* tmp = ValuesPtr;
        double* endTmp = EndPtr;

    return *std::min_element(tmp,endTmp);
    }


double MJArray::max() const
{
#ifdef RANGE_CHECKING
        if ( Size==0)
        {
            throw("cannot take max of empty array");
        }
#endif RANGE_CHECKING
        double* tmp = ValuesPtr;
        double* endTmp = EndPtr;

        return *std::max_element(tmp,endTmp);
    }


MJArray MJArray::apply(double f(double)) const
{
        MJArray result(size());

        std::transform(ValuesPtr,EndPtr,result.ValuesPtr,f);

        return result;
    }
```

The code here is quite straightforward. Some points to note: we only reallocate memory when the size becomes greater than the capacity so operator= and resize check size against capacity. This reduces the number of memory allocations necessary. The data member EndPtr is optional in that its value is determined by ValuesPtr and size. However, having a pointer for the start of the array and the end of the array leaves us very well placed to use the STL algorithms. These generally take in two (or more) iterators which point to the start of the sequence, and to the element after the end of a sequence, which is precisely what ValuesPtr and EndPtr respectively do.

We therefore use the STL algorithms to perform mundane tasks such as copying, taking the min and taking the max, and soon, rather than writing loops to do them ourselves. As well as saving us coding time, the general principle that we should use pre-defined routines rather than user-defined ones is a good one; pre-defined routines are generally close to optimal and we have the advantage that, as part of the standard library, another C++ programmer should recognize and understand them instantly.

Appendix D

The code

D.1 Using the code

The source code is downloadable from www.markjoshi.com/design. The code has been been placed in three directories: C/include, C/source, and C/main. Each main program indicates the source files that must be included in the same project for the code to link. The include files are included using < > so the directory C/include must be included in the list of places your compiler looks for include files. In Visual C++, the directories for include files can be changed via the menus tools, options, directories.

Makefiles, project files, etc. are not included as they are highly compiler dependent.

D.2 Compilers

The code has been tested under three compilers: MingW 2.95, Borland 5.5, and Visual C++ 6.0. The first two of these are available for free so you should have no trouble finding a compiler that the code works for. In addition, MingW is the Windows port of the GNU compiler, gcc, so the code should work with that compiler too. Visual C++ is not free but is popular in the City and the introductory version is not very expensive. In addition, I have strived to use only ANSI/ISO code so the code *should* work under any compiler. In any case, it does not use any cutting-edge language features so if it is not compatible with your compiler, fixing the problems should not be hard.

D.3 License

The code is released under an artistic license. This means that you can do what you like with it, provided that if you redistribute the source code you allow the receiver to do what they like with it too.

Appendix E

Glossary

anti-thetic sampling – a method of improving convergence in Monte Carlo simulations by following each sample by its negative.

class – a user-defined type.

constructor – a member function that has the same name as its class. It provides a way to create objects from the class.

container – a class with the main purpose of holding other objects.

decoration – the act of wrapping a class around another class in such a way that the interface does not change.

encapsulation – the process of representing a concept atomically in terms of a single class.

function – a routine inside a program to which information may be passed and/or returned.

inheritance – defining classes in such a way that they take on the attributes of an existing class plus additional characteristics.

iterator – a class that is similar to a pointer and, in particular, it can be incremented and dereferenced.

member function – a function associated with objects of a particular class.

method – another name for a *member function*.

object – a variable that comes from a class.

pattern – a code design.

pointer – a variable that points to a location in memory.

standard template library – a collection of header files with properties defined by the standard which provide a collection of container classes and algorithms.

STL – shorthand for *standard template library*.

template – a piece of code which is written to work with any class that defines certain chosen methods.

variable – a quantity that is stored within a program and can change in value.

wrapper – a smart pointer class that handles memory allocation and deallocation.

286

Bibliography

[1] A. Alexandrescu, *Modern C++ Design*, Addison-Wesley, 2001.

[2] M. Baxter & A. Rennie, *Financial Calculus*, Cambridge University Press, 1999.

[3] T. Björk, *Arbitrage Theory in Continuous Time*, Oxford University Press, 1998.

[4] S. Dalton, *Financial Applications using Excel Add-in Development in C/C++*, Second Edition, Wiley, 2007.

[5] B. Dupire, *Monte Carlo: Methodologies and Applications for Pricing and Risk Management*, Risk Books, 1998.

[6] G. Entsminger, *The Tao of Objects*, Hungry Minds Inc., 1995.

[7] E. Gamma, R. Helm, R. Johnson & J. Vlissides, *Design Patterns: Elements of Reusable Object-Oriented Software*, Addison-Wesley, 1995.

[8] G. Grimmett & D. Stirzaker, *Probability and Random Processes*, second edition, Oxford University Press, 1992.

[9] E. Haug, *The Complete Guide to Option Pricing Formulas*, Irwin Professional, 1997.

[10] J. Hull, *Options, Futures, and Other Derivatives*, fifth edition, Prentice Hall, 2002.

[11] P. Jäckel, *Monte Carlo Methods in Finance*, Wiley, 2002.

[12] N. Josuttis, *The C++ Standard Library*, Addison-Wesley, 1999.

[13] M. S. Joshi, *The Concepts and Practice of Mathematical Finance*, Cambridge University Press, 2003.

[14] I. Karatzas & S. Shreve, *Brownian Motion and Stochastic Calculus*, second edition, Berlin: Springer-Verlag, 1997

[15] I. Karatzas & S. Shreve, *Methods of Mathematical Finance*, Springer-Verlag, 1998.

[16] J. Lakos, *Large Scale C++ Software Design*, Addison–Wesley, 1996.

[17] A. L. Lewis, *Option Valuation under Stochastic Volatility*, Finance Press, 2001.

[18] S. Meyers, *Effective C++*, second edition, Addison-Wesley, 1997.

[19] S. Meyers, *More Effective C++*, Addison-Wesley, 1995.

[20] S. Meyers, *Effective STL*, Addison-Wesley, 2001.

[21] B. Moro, The full monte, *Risk* **8**(2), 1995, 53–57.

[22] R. Merton, *Continuous-Time Finance*, Blackwell, 1998.

[23] R. Merton, Option pricing when underlying stock returns are discontinuous, *Journal of Financial Economics* **3**, 1976, 125–144.

[24] T. Muldner, *C++ Programming: with Design Patterns Revealed*, Addison-Wesley, 2001.

[25] M. Musiela, M. Rutowski, *Martingale Methods in Financial Modelling*, Berlin: Springer-Verlag, 1997.

[26] B. Oksendal, *Stochastic Differential Equations*, Springer-Verlag, 1998.

[27] S. K. Park & K. W. Miller, Random number generators: good ones are hard to find, *Comm. ACM* **31**, 1988, 1192–1201.

[28] W. H. Press, S. A. Teutolsky, W. T. Vetterling & B. P. Flannery, *Numerical Recipes in C*, second edition, Cambridge University Press, 1992.

[29] L. C. G. Rogers, *Monte Carlo Valuation of American Options*. Preprint, University of Bath, 2001.

[30] A. Shalloway & J. R. Trott, *Design Patterns Explained: A New Perspective on Object-Oriented Design*, Addison-Wesley, 2001.

[31] B. Stroustrup, *The C++ Programming Language*, third edition, Addison–Wesley, 2000.

[32] H. Sutter, *Exceptional C++*, Addison–Wesley, 2000.

[33] H. Sutter, *More Exceptional C++*, Addison–Wesley 2001.

[34] H. Sutter, *Exceptional C++ style*, Addison–Wesley 2004.

[35] D. Vandevoorde and N. M. Josuttis, *C++ Templates: The Complete Guide*, Addison–Wesley, 2002.

Index